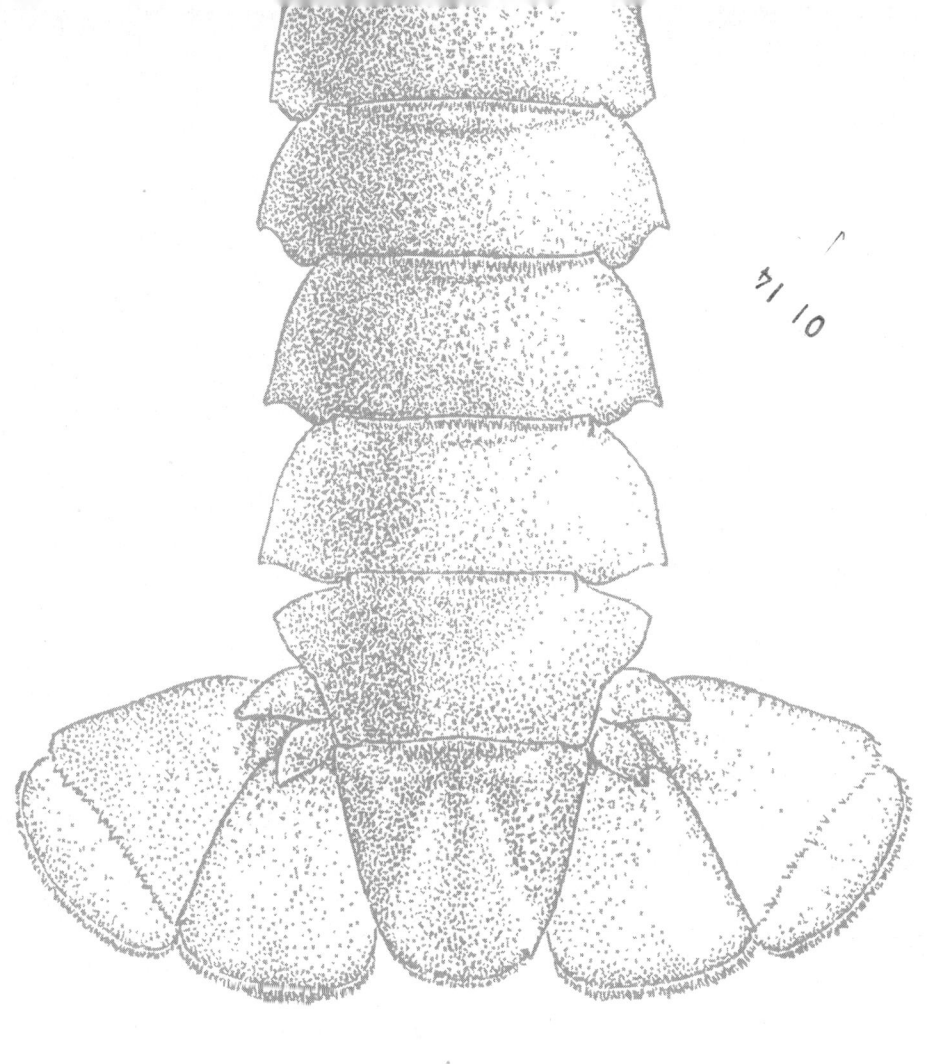

THE LOBSTER TRAP

THE LOBSTER TRAP

THE GLOBAL FIGHT FOR A
SEAFOOD ON THE BRINK

GREG MERCER

McCLELLAND & STEWART

McClelland & Stewart and colophon are registered trademarks of Penguin Random House Canada Limited.

The authorized representative in the EU for product safety and compliance is Penguin Random House Ireland, Morrison Chambers, 32 Nassau Street, Dublin D02 YH68, Ireland, https://eu-contact.penguin.ie

Library and Archives Canada Cataloguing in Publication data is available upon request.

ISBN: 978-0-7710-0632-6
ebook ISBN: 978-0-7710-0633-3

Photo insert: p. i, top: Nova Scotia Archives Photographic Collection: Blue Rocks, Lunenberg Co.: Harbour: Fisherman rowing lobster traps ashore; p. i, bottom: Library and Archives Canada/National Film Board of Canada fonds/e011175751; p. ii: courtesy of the author; p. iii, top: courtesy of the Gordon Parks Foundation/Library of Congress, Prints & Photographs Division, FSA/OWI Collection, LC-DIG-fsa-8d28609; bottom: Library of Congress, Prints and Photographs Division, NYWT&S Collection, LC-DIG-ppmsca-12782; p. iv: courtesy of the author; p. v: courtesy of the author; p. vi: top and middle: courtesy of the author; p. vi, bottom: Mason, Thomas Holmes, and Thomas H. Mason & Sons photographers. *Valentia: Confiscated Lobster Pots. Harbour in Background.* 1890/National Library of Ireland; p. vii: courtesy of the author; p. viii, top: Library and Archives Canada/Department of the Interior fonds/a047921; p. viii, bottom: Provincial Archives of New Brunswick.

Cover design by Andrew Roberts
Cover art: M/Adobe Stock
Typeset in Garamond 3 by Sean Tai

Printed in Canada

McClelland & Stewart
A division of Penguin Random House Canada
320 Front Street West, Suite 1400
Toronto, Ontario, M5V 3B6, Canada
penguinrandomhouse.ca

1 2 3 4 5   29 28 27 26 25

Penguin
Random House
McCLELLAND & STEWART

For Norah and Suzy,
my little lobsters

Contents

PART THREE: A Global Food

PART FOUR: Boom and Bust

THE LOBSTER TRAP

Prologue

Outside, on the boat's deck, I could hear them celebrating another successful haul. But I was busy inside, fighting a private battle all my own. I was on my knees, hugging the toilet, while the floor rose and fell around me and the whole room rattled from the droning in the engine room below.

I felt as if I was trapped in this tiny bathroom, bouncing inside a Canadian lobster boat, drifting alone on an angry ocean that tossed me around like a kid stuck on a violent carnival ride. Every time a big wave crashed over the bow, sending the vessel lurching, another swirl of nausea would cause me to empty my guts. My head was pounding, I had cold sweats, and my stomach ached from the constant retching.

It was then I had to remind myself of something—I *wanted* to do this. I wanted to experience what it was like to go lobster fishing in the weeks before Christmas, one of the most important seasons for the fishery. And I wanted to do it somewhere on the Bay of Fundy, a deep inlet of the Gulf of Maine that separates New Brunswick and Nova Scotia, where the weather can be notoriously awful at this time of year. Seasickness, I told myself, was part of the price of admission.

I grew up on Darlings Island, a small inland community not far from the New Brunswick seacoast, and I'd spent much of my life on the water. My parents were not fishermen—the preferred term for both women and men in the industry—but I'd been on plenty of boats and always felt at home around the sea. Nothing, however, had prepared me for the wild conditions on the big bay a day after a storm had kept fishing crews on land. All I could see outside the windows were the white caps of fierce waves, illuminated by the boat's headlights. The ocean felt big, black, and immeasurably powerful, a hostile wilderness that was endlessly dangerous.

Since I was a teenager, I'd been hearing about the boom in the lobster fishery and the spending spree it was fuelling in fishing villages up and down the coast. I was intrigued by fishermen who chose to chase the promise of riches on the wild ocean. I'd heard how the world had developed a taste for North American lobster, and how the boats couldn't catch them fast enough. How since the mid-1990s, Canada's harvest had more than doubled, to ninety-eight thousand tonnes a year—now more than triple America's output. But Canadian fishermen, nervously watching a second Donald Trump administration and its threats of crippling tariffs on imported foods from the U.S.'s northern neighbour, know how quickly things can change. What the Trump White House seemed to misunderstand, however, is that for an industry as integrated as the North American lobster trade, tariffs can be just as devastating for seafood producers on the U.S. side of the border as they are for Canadians.

While President Trump ultimately chose to hold off on tariffs against Canadian lobster, and Canada paused its plans for reciprocal tariffs, the threat of more trade wars and the uncertainty they bring continue to rattle the industry.

Most of the people I spoke to for this book were interviewed before China announced in March 2025 a 25 per cent tariff on Canadian agricultural and food products, including lobster, in retaliation for a levy on Chinese electric vehicles imposed by Ottawa. The $3.7-billion worth of tariffs piled even more pressure on an industry already grappling with a tariff fight with Washington, and threatened to devastate an overseas market that has become the second-largest buyer of live lobster in the world. Lobster is one of Canada's most valuable exports to China, with Canada sending $75 million worth of the shellfish to China in the month of January 2025 alone.

A growing trade war between the U.S. and China in the spring of 2025 further complicated Chinese consumers' love for North American lobster, at a time when American exports of the shellfish were still recovering from the effects of the last tariff fight between China and the U.S. in 2018. As it always has been with lobster, politics often gets in the way.

As a journalist who covered Atlantic Canada for local daily newspapers and later the national *Globe and Mail*, it was impossible to escape the impact of lobster. I spent years covering the fishery, writing about the wealth it was generating, and the heartache it brought when another fisherman was lost at sea. When the fight over a growing Indigenous lobster fishery exploded in Nova Scotia in 2020, with angry mobs ransacking warehouses being used by the province's Indigenous Mi'kmaq people, I reported from the wharfs and fishing boats that seemed braced for all-out war.

Over time, an idea began bouncing around in my head: I wanted to write about lobster in a bigger way, to better understand its evolution from a cheap protein for the poor into a global luxury food, and to explain how this iconic shellfish has pushed people to risk their lives, burn the boats of rivals, even try to kill those they suspect of stealing their catch. I wanted to understand the territorial tensions that exist in the industry, and why they push some fishermen to carry guns on their boats. Along the way, I hoped to peer into the psyche of fishermen who engage in one of the world's most dangerous jobs in order to bring us these strange creatures we see looking back at us in grocery store tanks.

As I began researching this book, I wanted to understand the transition happening in the fishery, and how the end of the lobster boom was affecting coastal communities, especially in the U.S., where in southern

New England the fishery has all but collapsed and is declining farther north in Maine. I wanted to learn what it means for fishermen when their traps start coming back empty, and what life after lobster looks like.

This is the story about the remarkable rise of the North American lobster fishery and the warning signs about its inevitable decline. It's a rapidly unfolding story as two powerful forces—ceaseless global demand and warming ocean temperatures—are putting pressure on lobster stocks in ways we've never seen before. The irony is that warming ocean water initially fuelled a boom in lobster populations in places such as the Gulf of Maine, causing optimal conditions for lobster production and driving the animals to migrate into areas that were traditionally too cold. But as sea temperatures continue to rise, past the sweet spot of lobster production of between 12 C to 18 C, that same patch of ocean floor becomes inhospitable. Lobsters are ectotherms, meaning their body temperature is regulated by the water temperature around them, so even the smallest changes can begin to alter their behaviour. As the seas warm, the old rules of lobstering—when, where, and how lobster should be caught—are being dramatically rewritten by environmental shifts no fisherman can control.

Telling this story led me across the globe, from Ireland, England, and France to South Korea and China, and across Canada and the U.S. Along the way, I spoke with dozens of fishermen, scientists, seafood exporters, and economists who are genuinely worried about the future of lobster. For many people who have earned a living from the sea, the changes taking place are deeply troubling.

Meanwhile, fishermen keep on fishing at a break-neck pace. Fishing communities don't need to be reminded of the dangers of single-minded pursuit of every last catchable piece of a species. We've seen this play out before. In Newfoundland and Labrador, fishermen watched their cod fishery collapse in the 1990s amid reckless overfishing, causing the province to lose more than forty thousand jobs almost overnight. Lobster, like all wild-caught seafood, is also a limited resource. Declining stocks and rising prices prompt more competition. For many fishermen who bought in during the boom years, this problem becomes a trap: they need to fish harder and harder for a shrinking harvest.

On the other side of the Atlantic, in the few places where European lobster are still caught, it's almost exclusively an export-only ultra-luxury food, saved for diners in fine restaurants or flown overseas to places like China and resorts in Saudi Arabia. Is this the future for the North American lobster, once so abundant that McDonald's put it into a cheap fast-food sandwich? And what lessons can we learn from Europe, where for centuries lobster was caught with little regulation or management?

I had to wonder if the era of affordable lobster was over as I visited places like Viking Wharf, an all-you-can-eat lobster buffet inside a sprawling shopping mall in Seoul, South Korea. There, people pay $100 U.S. to gorge themselves on red lobster tails and claws, stacked neatly on long trays of ice, then down bottles of lobster-branded sauvignon blanc and pose for a photo with a giant cartoonish lobster in a top hat outside the front door. How sustainable, I wanted to know, is this for a seafood that so far has proved ill suited for farming and can be caught only in one small corner of the North Atlantic?

These were the questions I was troubled by as, a few months into writing this book, I stepped into Taste of China Seafood Restaurant in Toronto's Chinatown on a cold and grey February afternoon. This eatery on Dundas Street West has been serving up seafood since 1997, and is known in the neighbourhood as the kind of late-night place favoured by the after-bar crowd and chefs looking for authentic Cantonese-style cooking.

Shortly after I took a seat at a round table in the corner, Mei Cheung emerged from the back and handed me a plastic-covered menu. It had hundreds of dishes, a blur of sautéed shrimp and crab, rice dishes, and black bean sauces, all printed neatly in a tiny font. The list seemed to go on forever. I didn't bother to finish reading it, and told Mei what

I was here for: lobster. She asked if I wanted it Hong Kong style, and I nodded.

Where I grew up, whole live lobster was almost always served one way: It came steamed or boiled, with a side of melted butter and nothing else. You had to crack the shell yourself and dig out forkfuls of the sweet, tender meat. It was always messy, with lobster juice flying everywhere, while the butter ran down your chin. As a boy, I always wondered how such a simple meal could taste so good.

At the turn of the last century, Chinese immigrants in places like Toronto, New York, and Vancouver and in cities and towns around North America began elevating lobster from the humble pot into a truly international dish. They took this traditionally East Coast food and made it all their own, adding flavours and Cantonese cooking techniques you'd never see at a lobster boil. They even invented something called lobster sauce, a whitish thickened seafood sauce that, ironically, contains no lobster—a cost-saving measure for newcomers who eventually replaced the shellfish with cheaper shrimp as the price of lobster rose beyond their reach.

I was thinking about all this as Mei dipped an arm into one of the bubbling tanks stacked against the wall, picked out a squirming lobster from among the live tilapia and sea bass, and waved for me to follow her into the kitchen in the back. One of the cooks hacked my lobster to pieces with four or five violent whacks of a large cleaver. The lobster was then doused in cornstarch batter and tossed into a large black wok. Chef Ping Yeung cooked it quickly, adding Chinese cooking wine, minced pork, garlic, dried fish, soy sauce, MSG, and sugar. Originally from Guangdong Province, on the South China Sea, he'd been cooking seafood nearly all his life. Ping looked like he could handle a flaming wok in his sleep.

In the span of about four minutes, my Nova Scotia lobster was transformed into a gloriously fried pile of chopped shellfish. Everything on the plate was delicious. Mei hovered nearby, repeatedly topping up my green tea. In Toronto's Chinatown, she told me, this used to be the meal you ordered when you had something to celebrate. But lately, fewer and fewer people were ordering it when they came into the restaurant.

Lobster is too expensive now, she explained, and the younger crowd drawn to Chinatown's flashing lights want cheaper meals like noodles or barbecue. When Taste of China opened in 1997, just as lobster catches in North America were beginning to soar, Canadian fishermen were being paid around four dollars a pound for their lobster. By the winter of 2024, those same fishermen were being paid eighteen dollars a pound. Taste of China has to keep raising their prices to afford to keep lobster on the menu. And it appears it's reaching a tipping point.

"There used to be a lot more Chinese restaurants selling lobster this way," she said, looking out the window. "But every year there's less and less."

Mei was talking about customers, not the supply of lobster. But if you sit in her restaurant and squint and imagine the place lobsters come from, two days' drive eastward where North America crashes into the Atlantic Ocean, you'll hear more and more fishermen saying exactly the same thing: Every year there's less and less.

PART ONE

A WARNING

1

The Great Lobster Machine

HALIFAX, Nova Scotia – Karl Riches likes to come here at night, when the big, noisy fans are turned off and the tidy stacks of polystyrene boxes seem to be conducting their own strange symphony. It is the unsettling sound of tens of thousands of live lobsters scratching against their containers, trying to snap their fellow captives' claws off and escape, yet constantly foiled by the rubber bands that keep them from killing and eating each other.

For thousands of people whose livelihood depends on the North American lobster fishery, the eerie noise inside this cavernous chilled warehouse near the Halifax international airport is also the sound of money. Less than a day earlier, these creatures were being kept alive in saltwater pounds and indoor holding tanks around Canada's Maritime provinces. By tomorrow, they'll be squirming in markets and restaurants and kitchens in Asia and Europe, where they have become a luxury food for affluent consumers who seem to have an unending appetite for them.

Inside Gateway Facilities' cavernous complex next to the highway, Atlantic lobsters are unloaded off fleets of trucks and quickly packed for international flights. They're precious cargo, handled as gently as eggs. Karl, a former Royal Air Force logistics manager who speed walks around the tarmac pointing out the parking spaces for Boeing 747s, is Gateway's general manager, and he knows there's no time to waste. In a busy month, Gateway will load more than sixty planes full of live lobsters, some of them carrying as much as 100 tonnes each.

Lobsters are notoriously fickle travellers, so in each box, a small temperature sensor constantly spits out readings, sending alerts to managers' phones if things get too warm or too cold. At every step of the process, the lobsters must be kept at 4 degrees Celsius. Before this place existed, most of the lobster coming out of Atlantic Canada had to be flash-frozen or loaded live onto trucks and shipped on the long ribbon of asphalt to Montreal, Toronto, or Boston, where, if they survived that trip, they could only then be packed onto planes for overseas markets. Some 80 per cent of the lobster leaving this specialized warehouse—the only one like it north of Miami—will end up in China, about a nineteen-hour flight away. In a job where delays come with steep price tags, every minute counts.

In 2011, Gateway shipped about 4,800 tonnes of live lobsters around the world. Ten years later, it was shipping six times that much. The orders keep coming in, and the planes keep going out. South Korea. Singapore. Beijing. Qatar. Paris.

"The growth has been incredible," says Karl. "I don't want it to stop."

But even if he wanted it to, *it can't stop*. The miracle of modern logistics has created a supply chain that can deliver lobster from a wharf in Yarmouth, Nova Scotia, to the Jingshen Seafood Market, China's largest, in the span of a weekend. Gateway is an important entry point in a system built to consume millions of pounds of lobster every year. There's too much riding on it to stop. The mass-market commercial fishery is designed to pull as many lobsters off the bottom of the ocean floor as the market will bear, whether or not the species can sustain it.

Only the invention of refrigerator trucks, state-of-the-art cargo planes, and on-demand logistics has allowed people on the other side of the planet to look into the beady eyes of a live lobster from Nova

Scotia, or Maine, or Prince Edward Island, and then have it for dinner. Until a few decades ago, if you wanted to buy lobster in a place far from where they are caught, it arrived precooked in a tin, or frozen in the grocery's seafood aisle. In past eras, businessmen trying to do what Gateway does—bring live lobster to the world—attempted this by train or boat and learned the hard way that once lobster are taken out of the sea and crowded together, the countdown to death begins.

There have been many failed efforts to transplant live North American lobster to the Pacific Ocean and other parts of the world in vain attempts to establish new fisheries there, but in every case, these creatures have stubbornly refused to survive anywhere else. They are so infinitely particular about their habitat and the temperature of the water, and so prone to predators and disease, that the only place on the planet they seem able to thrive, outside of a laboratory setting, is the coast of New England and Atlantic Canada.

There are many types of lobster in the world, but this book is focused squarely on the species that is by far the most commercially important, *Homarus americanus*. It goes by a number of names, thanks to the marketing efforts of seafood companies—American lobster, Atlantic lobster, Boston lobster, Maine lobster, and Canadian lobster. It is all the same crustacean, recognizable for its large, shrimp-like body and ten legs, including two meaty claws designed for pulverizing shells and tearing soft flesh. Between the northeastern states of the U.S. and Canada's eastern provinces, where lobster roam the rocky seabed, they're worth about $7 billion a year to local economies.

One of the first records of Europeans encountering North American lobster came in 1607, when Robert Davies, captain of a four-hundred-ton ship called the *Mary and John*, described the spectacle in his journal. His ship, which sailed between England and the American colonies four times, was carrying colonists to the Popham Colony, a new settlement in Maine near the Kennebec River. This settlement turned out to be an abject failure, abandoned after just fourteen months, following the deaths of most of its inhabitants.

But a discovery made en route to their short-lived New World home would prove far more important to the economic development of the northeastern corner of North America. On August 2, 1607, the vessel had anchored at the mouth of the LaHave River in southern Nova Scotia while its crew began collecting fresh water from a nearby island. A group of Mi'kmaq women and children approached the ship, attempting to trade glass beads, as they had done with French explorers before, but the English sailors had little interest, Robert wrote. They wanted food.

That evening, when some of the ship's crew rowed out to the island where previously they had found berries, they noticed the shallow water below them was alive with lobsters. One of the men fastened a long hook to a wooden pole and began plucking the creatures from the water one by one. Soon their little boat was brimming. "The boat," Captain Davies wrote, "went presently from the ship unto a point of an island and there at low water in an hour killed near 50 great lobsters. You shall see them where they lie in cold water not past a yard deep and with a great hook made fast to a staff you shall hitch them up. There are a great store of them you may near load a ship with them, and they are of great bigness. I have not seen the like in England."

Despite their abundance, however, it would be several centuries before most Europeans could enjoy North American lobster. Unlike fish, which can be smoked, salted, or dried for the long voyage across the ocean, lobster needed to be cooked within two days of leaving salt water. They were a staple diet of Indigenous Peoples in the New World's coastal region, but for most Europeans, lobster was an expensive food reserved for the rich. While it may have been the first time these English sailors had seen an American lobster, they would have been familiar with its European cousin, the slightly smaller, blueish-tinged *Homarus gammarus*, the European or common lobster. If they were clever enough to catch it, and brave enough to cook it and crack open its formidable shell, they were rewarded with a delicate and sweet meat unlike anything else in the sea. Historian Joseph Gough, who wrote frequently about the Canadian fishery, suggested it took some bravery to initially see lobster as food, saying "The first person to eat a lobster must have had great nerve, to hope for anything good behind the claws, spidery legs, and bulging eyes. But people gradually learned."

They certainly did. Eventually, these new visitors realized the value of a lobster population so plentiful they would commonly blanket the beaches after powerful storms. They began to develop new ways to catch, cook, preserve, pack, and ship lobster. Once they started, they couldn't stop. Lobster evolved into an ever-growing seafood industry that has brought untold riches to rural coastlines.

Our four-hundred-year fascination with North American lobster has also had an irreversible impact on the species. The jumbo lobsters caught in colonial times, specimens reaching four feet long and weighing as much as forty-five pounds, are never seen anymore. Instead, most of the lobster shipped worldwide by companies such as Gateway are in the one- to three-pound range—partly because that's what consumers want and partly because our intense overfishing has actually changed lobsters' life cycle.

Marine biologists who study these creatures say female lobsters are reaching sexual maturity at a younger and younger age. That's the product of years of intensive selective fishing, which saw fishermen harvesting females before they'd had a chance to reproduce. In most places that fish Atlantic lobster today, females found to be carrying eggs— "berried," in fishermen's parlance—must have their tails "V-notched," or marked with special clippers to indicate to other lobstermen they must be released. That conservation measure has helped fishermen enjoy record-level catches by ensuring that juvenile lobsters get a chance to reach maturity.

What concerns biologists, however, is the growing number of females hauled up from the bottom with fewer eggs tucked on the inside of their tails. It's an important biological indicator of a species going through a dramatic change. Since 2012, annual surveys that count the juvenile lobsters resettling into rocky, coastal nursery grounds in New England and Canada have shown declining numbers, meaning the future of the fishery is going to be one of diminishing returns.

There are other troubling signs that the era of peak lobster is already over in many of North America's traditional lobstering regions. In Maine, the heart of the American lobster fishery, fishermen caught less than eighty-six million pounds of lobster in 2024—a fifteen-year low, according to state records. Overall, catches have been trending downward for years, despite historically high prices that made the shellfish worth $890 million U.S. to Maine in 2021. This drop-off has come after several decades of record-setting catches, starting in 1990, when lobster landings in the Gulf of Maine broke a record that had stood since 1889.

There are warnings in Canada, too. The 2024 fishing season in the Northumberland Strait, the body of water between New Brunswick and Prince Edward Island that has traditionally been some of the most productive lobster fishing grounds in the country, began abysmally. Fishermen reported their landings—the industry term for the lobster harvest measured once back onshore—were way down, by as much as 40 per cent, from average. The theory is that warming water temperatures, trending upward for years, had finally tipped to a point where the lobster had become lethargic, shedding their shells months later than normal, and less likely to seek out the bait in fishermen's traps.

"So far, it's a season we'd like to forget," moaned Luc LeBlanc, an advisor for the Maritime Fishermen's Union. "The water in the Northumberland Strait is at historic temperatures. It's very, very hot." Down south, in places like Connecticut and Cape Cod, the lobster have all but disappeared. This disruption of historical norms in the fishery is only accelerating as climate change displaces fishermen and alters coastal communities in permanent ways.

Meanwhile, in the lobster's most northern habitat range, on Quebec's eastern shore and in Newfoundland and Labrador, fishermen are reporting a lobster boom unlike anything they've seen before, one that has the potential to repair the economic devastation to their communities that still lingers from the collapse of the cod fishery in the 1990s. But even in these northern regions, this lobster fishing frenzy has downsides. On the Magdalen Islands, an archipelago in the Gulf of St. Lawrence a five-hour ferry ride north of Prince Edward Island, traditional herring smokehouses complain they can't get fish anymore because it's all going to bait for the lobster industry.

Modern lobster's boom-then-bust phenomenon is a problem on both sides of the Atlantic Ocean, where lobster fishermen are going farther and farther out to sea, and putting in longer hours on the water, to meet the unending demand for their shellfish.

"There are a lot of signs the lobster population is under intense stress," said Bryce Stewart, an affable Australian biologist who studies European lobster stocks along the Scottish and English coasts.

For decades, British fishermen hauled up lobster without an accurate system to measure how much they were taking from the sea, complicating conservation efforts. Today, those fishermen are fishing in deeper waters, using more pots, and upgrading their boats so they can reach previously unexploited lobster grounds. A few generations ago, they were still fishing in the old style of their grandfathers—launching wooden boats from long, flat beaches, and hauling their catches in by horse.

If you travel to the English seaside in Scarborough, a city popular with holidaying Britons for its soaring cliffs, sandy beach, and busy, touristy waterfront, British-caught lobster is almost impossible to find. A few restaurants serve it, but for the most part, what lobster is left in local waters is being shipped across the English Channel to restaurants in France, Belgium, Spain, and Italy, where diners are more willing to pay a premium for it. As I wandered among the blinking lights along Scarborough's crowded beachfront, people looked at me strangely when I asked them if they had any local lobster on the menu.

"Lobster? Sorry, love, it's too expensive," said the owner of a seafood takeout stand that sells prawn sandwiches and steamed crab by the Scarborough beach.

As the industry faces these duelling pressures—declining catches in its traditional fishing regions and rising prices fuelled by global demand—some worry lobster is destined to become a luxury food that only the most affluent of us can afford to eat. They argue convincingly that's already happened. The golden age of cheap lobster is clearly over, as restaurant chains quietly remove it from their menus and food processors switch to cheaper alternatives. As the price of lobster rises, it's spawning not just a gold rush mentality on the water, but a new wave of violence between those fighting over who has the right to catch it.

With no limits on the catch—fisheries regulators in both Canada and the U.S. restrict the number of traps fishermen can use, but not the amount of shellfish they can haul up—we've built a fishery dependent on the superabundance of lobster. In Canada, the total catch is twice what it was fifteen years ago and has increased tenfold in some areas. Fishermen, many up to their necks in debt because of the high cost of entering the fishery and with no choice but to keep fishing, are going out in worse weather and taking risks with their own lives because the world's demand for lobster is unending.

The lobstering boom has also killed diversity in the fisheries, making it the only species that really matters to fishermen in many parts of New England and Atlantic Canada, and by far the most valuable catch in most northeastern seagoing communities. Combined with the influence of global seafood corporations, and the demand for lobster across the globe despite its rising price, government fishery managers are under intense pressure to maintain a status quo despite signs they need to slow down the harvest.

Add to all this the long-running problem of off-the-books cash sales, which make it difficult to get an accurate assessment of how many lobster are really being caught in a given year. In the lobster fishery, fishermen operate on what's essentially an honour system to report how much lobster they're catching, as a condition of their licence. There's simply too many boats, and too many wharves, for government fisheries officers to be able to monitor harvests effectively.

Some estimates put the size of this black market in Canada at between 10 to 30 per cent of the total catch. And as the value of lobster has grown, so have the income tax implications for those catching it—making some fishermen increasingly incentivized to under-report the size of their catches. Given the competition for lobster among processors, there's no shortage of seafood buyers willing to pay in cash, too. That's a big problem for a species that needs good data to help guide conservation measures, now more than ever.

"A lot of people would rather just continue to be in denial, which is what we've done about it for the longest time," Nat Richard, executive director of Canada's Lobster Processors Association, told me. "Obviously, it's difficult to talk very openly about. And I have no illusions about the challenges of implementing dockside monitoring across such a large fishery. But at the same time, we can't continue to turn a blind eye to this problem."

If the North American lobster industry wants to have its best shot at a reliable future, it needs to get control over this dark corner of the fishery, he says. Increasingly, consumers around the world want to know their food is being caught in way that minimizes environmental impact and ensures healthy seafood populations. It's especially important at a time when warming oceans are rapidly changing marine ecosystems, and more than a third of fish stocks are already being fished at unsustainable levels.

Seafood buyers turn to organizations like the Marine Stewardship Council (MSC), a non-profit organization that sets standards for sustainable fishing, to help them feel good about what they're eating. Nat worries these kind of MSC certifications can be at risk if part of the lobster industry remains outside of monitoring measures that protect against overfishing.

"There is a portion of lobster from the commercial fishery that is not being reported, which, frankly, is a condition of license for any harvester," Nat says. "And when we know that is happening to the degree that it is happening, what does that say about our ability to manage this resource with solid data, and having an accurate read on what's being fished year to year?"

2

The Last of the Hunter-Gatherers

DIPPER HARBOUR, New Brunswick – The lights of the harbour are fading in the distance as a voice crackles over the radio. "What's the weather like?" it asks.

It's two hours before sunrise, and Brad Small is steaming his lobster boat, *Small Fortune's*, straight out into the blackness of the Bay of Fundy, where a southwesterly wind is blowing at a steady twenty-five knots, or just under thirty miles an hour, hard enough to create big, heavy swells that toss his nearly fifty-foot vessel about like a toy. Headlights at the top of the boat illuminate the white caps of waves swirling in all directions, but beyond that, everything is dark. The temperature is −4 Celsius, a biting cold that, combined with windchill, numbs your fingers and stings your face. Every few minutes, a rogue wave crashes violently into the side of the boat, forcing the crew to brace themselves against the walls to keep from falling down.

"Just awesome," Brad deadpans, without hesitation.

For many Canadian fishermen, this is *lobstering* weather. Late November is when there is money to be made out here, often on rough and unwelcoming seas, as an invisible army of lobster begin their annual march away from the shallow, warmer coastal waters where they spawn and shed their shells before heading back to deeper water. It's on this yearly migration route that Brad and his crew have laid their traps—three hundred coated wire boxes that are baited with a pungent mix of sardine heads and tails, chopped-up redfish, and crab. On a good day, early in the season, they'll haul up three thousand pounds or more of squirming lobster. But within a month or so, this feeding frenzy is all over. The ancient instincts that dictate so much of lobster life tell them to stop feeding, and suddenly Brad's traps will be empty. Shortly after Christmas, most fishermen along this coast will put their traps away until spring, when the lobster start moving and are once again looking for food.

"It's like turning off a light switch. They just stop," says Brad, who at sixty-two is one of the more experienced and respected fishermen out here.

His first mate is Tom Duke, thirty-two, who with his bushy black beard and navy blue tuque looks exactly like you'd expect a lobsterman to look. Two deckhands are Howard Robbins, a fifty-three-year-old from Lubec, Maine, with a thick New England accent that turns "harbour" into "hahbah," and Ethan Lomax, a wiry twenty-six-year-old who consumes energy drinks and cigarettes with equal enthusiasm, sometimes whooping in excitement at nothing in particular. Tom, who gets paid based on a share of the catch each season, directs the other two on deck while Brad, the captain, runs the boat, using a GPS plotter and digital mapping software to track down his buoys hidden around the bay.

The four seem unfazed by weather that would cause most people to lose their lunch. Seasickness is not a problem for those who have had their sea legs since they were little kids. Everyone on board learned how to fish from their fathers or grandfathers, and they work methodically and quickly, the product of years of repetition. The boat they travel on is designed for rough weather, and is far more stable than the vessels Brad started on in the 1970s. It's built for scooping up huge catches of scallops in heavy winter seas, sitting wide and close to the ocean's surface, with large ballast holds designed to take on water to stabilize the boat. It has a full kitchen, and bunks below decks for five people. Every door and all the appliances have latches to keep them from flying open when the boat lurches, which it does often.

Brad knows plenty of fishermen have died pushing their luck against an ocean far more powerful than they are. As he talks, he's always scanning the sea, squinting into the horizon. Big swells crash over the bow as the boat plows through the sea and then dips its nose down into the next wave. While they don't fear heavy weather like this, most lobstermen also understand there are limits to their boats—and their bodies. After his brother had a heart attack on the water, nearly three hours out from the harbour, Brad keeps a bottle of Aspirin on the dashboard. He knows accidents can happen quickly. Tom once got his boot tangled in a trawl line and was dragged off the end of the boat, slipping loose seconds before he would have been pulled to the bottom of the bay. Howie was working on another boat a few years earlier that was hit broadside by a big wave and flipped upside down. Another time, he watched a deckhand break his leg when it was caught up in a winch. Most lobstermen who've worked at sea long enough have stories like these. This job comes with risks, and Brad admits it's on his mind whenever he's on the water. He feels responsible for the safety of the men who work for him—"All the time," he says, and then he goes quiet.

The first set of twenty traps, strung together in a line called a trawl, is anchored an hour's ride out into the ocean in about 60 fathoms, or 360 feet, of water. It's marked by a large "balloon," a round, orange buoy that indicates the location of the trawl. Smaller black-and-white buoys tell other fishermen these traps belong to the *Small Fortune's*. The fishing line itself is another marker, used to identify the country, fishing zone, and fishery the gear came from if it's found drifting at sea or tangled up with a whale.

Brad eases back on the engines and it's time to get to work. Howard grabs the balloon with a long wooden gaff, a pole with a metal hook on the end, and pulls it on board, before looping the attached line into the boat's hydraulic hauler. The machine is flicked on and squeaks loudly as it begins hauling the heavy line of traps up from the seafloor, starting with the trawl's anchor. It takes a few moments for the first trap to appear, and all eyes on deck are trained on it as it emerges from the water. Three squirming lobsters are inside, and the crew immediately goes to work. Howard places the seventy-five-pound trap onto a metal rack at the edge of the boat, and Tom swiftly pops it open, grabs the lobsters with a gloved hand, and places them in a nearby plastic bin, where Ethan will measure them for legal size with a hand tool, placing rubber bands on their claws if they're keepers. If they're too small or found to be bearing eggs, he throws them over his shoulder back into the sea. Howard removes what's left of the old bait bag and places a new, foul-smelling one in the trap. Within about fifteen seconds, it's ready to be stacked on the deck, to be returned to the sea when the trawl is all done. All the while, the hauler keeps loudly pulling more traps up from the water. Any delay on deck and they'll have a logjam waiting to be cleared.

"Time is money," Tom says, as he braces himself against the boat's rocking. "Every delay costs fuel, and costs us money."

Inside the cockpit, Brad keeps an eye on the bank of screens that indicate depth, tides, and the location of other boats on the bay, marked by blinking orange dots. Nearby are gauges monitoring oil levels, diesel consumption, and the bilge pumps taking on seawater to stabilize the boat against the heavy waves. Echo sounders transmit a sound pulse down into the water as the boat cuts through the waves; it bounces back up when it hits fish or something solid, giving the captain an idea of what's below. But in today's modern vessel, so much is automated. When Brad began fishing, storms could catch fishermen off guard and blow them far off course. He once spent two days in Digby, Nova Scotia, waiting for a storm to blow through after the weather pushed his boat all the way across the Bay of Fundy. Today's modern vessels can detect bad weather long before it arrives. Brad stands as he operates the boat, out of habit, but he could almost run the *Small Fortune's* on autopilot from the plushy captain's chair. He tells me, almost wistfully, that the old paper charts he learned on have long been replaced by software that takes the guesswork out of running a fishing boat. In those days, fishermen relied on landmarks and compasses to find their way. On this morning, tracking down buoys in the dark requires only that a captain click on the location he wants to go to, and the boat makes a direct line for that target.

"Today, a kid with a Nintendo could do this," he says.

November is a critical time in the Canadian lobster fishery. It's when lobster buyers and exporters are trying to fill Christmas orders from restaurants, grocery chains, and processors around the world. On this morning, the wind kept some crews home, but Brad and his crew, having been onshore for three days, know their traps will be full. They set out at five. It will be dark when the boat returns to harbour that evening.

When you're handling a live shellfish that can survive out of water only so long, time obviously matters. But fishermen in both Canada and the U.S. are also in a larger race against time, as the lobster boom of a decade earlier fades from view and they rush to catch as many of the shellfish as they can before the population suffers more serious declines. Already around the bay, catches are declining. Down the coast, fishermen are talking quietly about a 30 per cent decline in the harvest in the first few weeks of the season. Today, the *Small Fortune's* will haul in about 1,400 pounds of lobster, just shy of its mid-season target. But it's still a good day on the water. Back in Dipper Harbour, lobster buyer Lyman Crawford is paying $9.50 a pound for their catch.

When the boat returns to the wharf, it's past six in the evening and low tide. There's barely enough water for Brad to pull up to the "lobster car," a floating barge where young men unload crates of lobsters and place them in a hoist that delivers the shellfish to a waiting truck, two storeys above on the wharf. The lobsters from the *Small Fortune's* will be headed to the U.S. before the night is over, most of them loaded onto planes at the Boston international airport and flown around the world. This interconnected global system for shipping lobster has been good to fishermen like Brad, and especially to buyers like Lyman. He says seasickness kept him on land, so he got into the export business. Now he drives a Lamborghini.

Most of the seafood consumed around the world today comes from fish farms, the kind that Ethan works on when he's not lobstering—typically giant pens close to shore where tens of thousands of fish, crab, and shrimp are raised at a time in controlled environments. They're constantly monitored for health, fed a steady diet of antibiotics and processed feed, and harvested just when they reach the ideal market weight. Those animals keep few secrets. Their entire lives, from the hatchery to the supermarket, are already plotted out on a spreadsheet before they're born. The workers who run these farms can get to their jobsite in small boats not designed for the open seas, and work predictable shifts, like clocking in at a factory.

Lobstering is not like this. Unlike fish farmers, lobstermen can't raise their catches. They can't grow more of them, no matter how much the world may be willing to pay. Instead, they must go to where the lobster are, often in weather and sea conditions most sane people would avoid. And while the fishery drops untold tonnes of bait into lobster grounds every season, providing a steady supply of food, most of the lobster's life is not under their control. Lobstermen are at the mercy of both the oceans and the mysteries of population swings beyond our understanding. Until they're finally caught, lobster aren't penned in by cages, and can travel hundreds of miles along invisible migratory routes. The ocean bottom they move along is often full of hidden ledges, canyons, and crevices unseen by the fishermen working up on the surface, where the sea surrounding a boat appears to be an endless vista of sameness in all directions. No matter how much experience they have, lobstermen are constantly being humbled. What worked last season, or even last week, may not work this season or week. In that sense, they're the last of the hunter-gatherers. Man versus wild, but with better technology than their ancestors.

"People tell me there are easier ways to make a living," confesses Tom, in a brief moment of pause while the boat steams toward the next string of traps. "But this gets in your blood. It's almost like an addiction. When the season ends, you're already thinking about when you can get back out there. As it gets closer and closer to opening day, it's all you can think about."

Like Brad, Tom comes from a family of fishermen who were raised on the sea. But the world of today's lobsterman is markedly different from even a generation ago. The money is better, but there's a feeling in the industry that this may be the last great gold rush before a long and troubling decline.

People have fished out of Dipper Harbour, a small indent in the jagged New Brunswick coastline, since it was first settled, in 1786, by a Scottish soldier named Hugh Campbell. Hugh was given land by the British Crown after fighting for the king during the American Revolutionary War, and he learned to work with the tides to build deep-water weirs, circular nets suspended on posts used to catch herring. Centuries before there was ever a settlement here, this area was a source of food for Indigenous tribes who caught salmon, mussels, and other shellfish. The land was always rocky and poor for farming, but the sea provided. When the community's only church was built, a little white-sided and black-roofed building just steps from the main road, they named it, naturally, after Saint Brendan, the patron saint of seafarers. The villagers paid for it by hosting dances on the wharf.

In the eighteenth and nineteenth centuries, lobster fishing was done on a very small scale, typically by one or two men in a hand-rowed dory or sailboat, perhaps hauling up five wooden pots at a time, and always working close to shore. It was subsistence fishing. Long before the invention of commercial canneries, fishermen and their wives would process their catches in their own homes, boiling the meat in large pots on the stove. They would pack the lobster meat themselves, using lead solder to seal the cans. It was a crude method, prone to spoilage, but without refrigeration or modern transportation networks, there were few other ways to get lobster to buyers who didn't live nearby. Eventually, boats with live wells, or holds filled with seawater, were built to take live lobster to cities farther away on the east coast and down into New England, but even so, lobster remained for the most part a regional and part-time industry.

Today, with its old century homes, clotheslines flapping in the wind, and moss-covered cemeteries, Dipper Harbour might look like it belongs in a time capsule. But the fishermen who work here now are part of a fishery that would be unrecognizable to their ancestors, the critical first link in a sophisticated global network that can deliver live lobster from the bottom of the Bay of Fundy to a sidewalk seafood restaurant in Qingdao, China, in the span of a few days.

Even just a generation ago, lobstering was a remarkably different game than it is now. In Brad's father's day, fishermen who fished too hard were considered greedy by their neighbours. They took Sundays off, and left their boats at the wharf when it was time to pick cranberries or go hunting. That culture has completely changed, Brad says. Now, the goal is to catch as many lobsters as you can, always pushing harder and harder to squeeze as much revenue out of your boat as possible. That's partly driven by the money involved and the debt loads carried by many fishermen. Brad bought his first boat, after graduating high school in the late 1970s, for $4,800. It was small and made of wood, and could be blown far off course in bad weather. The modern vessel he uses today costs more than $1 million. It's like going from a horse-drawn carriage to a Cadillac in the span of your lifetime. Add the price of modern boats to the soaring costs of bait, fuel, and bank loans, and today's lobsterman is under a level of financial pressure that his father and grandfather never experienced.

The problem is that North America's lobster industry is now on the downside of a lobster boom that caused a lot of people to borrow heavily to enter the industry, paying obscene prices for licences, boats, and equipment. While the fisheries in Canada and the U.S. both work under the principle of owner-operators, there are important differences. In Canada, it's a limited entry industry, which means you can only legally catch lobster by buying someone else's licence—prized access that costs as much as $1 million in recent years. In the U.S., anyone can enter the fishery as long as they buy a licence from the state. But on both sides of the border, during peak lobster years fishermen bought bigger boats and more expensive equipment, and as catches have declined and costs have risen, many are overextended and can't afford to catch fewer lobster. For some of them, drowning in debt and worried about bills coming due, conservation of the resource is an afterthought.

"When you go out, you know the bank lender is always waiting on the wharf, so you'd better catch as many lobster as you can," Brad explains, while sitting on the deck of the *Small Fortune's* when bad weather has kept the boat at dock. "That's why a lot of guys who got into lobster during the boom years, they're not worried about the next generation. They're only fishing for today."

Just as he says this, Richard Thompson, whose family company Coastal Enterprises has been buying and processing seafood in the area since the 1970s, hollers down from the open window of his white Toyota pickup truck. He's parked up on the wharf, which at low tide sits well above the deck of Brad's boat. He likes to come down to the water regularly to check in on catches, pulling on a cigarette while he talks to fishermen he's been buying from for years. Like Brad, he's also worried about where the industry is going.

"It helps to make an old man older," he says, pushing his dog Rufus out of the passenger seat to give me a break from the howling wind outside. Rufus, a caramel-coloured golden retriever, grumbles and sticks a wet nose in my ear as I climb in.

When I ask him how things have changed, Richard just waves his hand at all the new pickup trucks crowding the wharf's edge. "I mean, the trucks tell you a lot, right?" he says. "The trucks that are parked on the wharf, they're not the same trucks that were parked on the wharf fifty years ago. You know, they were old junkers back then, they had no mufflers, but they were always the trucks that tended to the wharf. And now there are $125,000 crew cabs with leather seats for their dogs," he says, shooting a side-eye at his pet.

Fishermen here have been able to afford top-of-the-line trucks because of a historic rise in both lobster populations and global demand for the shellfish. Lobster catches in the Bay of Fundy grew from less than 500 tonnes in the 1970s to more than five times that in 1999. By 2020, fishermen on the bay were catching more than 9,200 tonnes a year. The money that has flooded into the fishery in step with those jaw-dropping harvests has completely altered coastal communities. It's brought a new generation of fishermen who bought in at a time of ever-growing catches, with bigger, more expensive boats, and no memories of scarcity.

"Now all of a sudden, the dollar bills are showing up and everybody wants a piece of it," Richard says. "All this is pressure on the lobsters. Whether or not the stocks will sustain, you know, what is happening here now in the industry, I guess we'll have to wait and see. We have to put our faith in science, and hope that we may be okay."

What has people like Richard worried is that, up and down the coast, catches are suddenly declining far more quickly than the lobster boom arrived. In 2023, fishermen around the bay were grumbling about a sharp drop in their harvests, some reporting decreases of as much as 40 per cent. By January 2024, concern over the health of lobster stocks in the U.S. resulted in a new restriction on the minimum size of lobster that can be legally caught. The measure, which was triggered by scientific surveys in the Gulf of Maine that showed a continued decline in the numbers of young lobster, could mean that 20 to 30 per cent of lobster being caught are suddenly off-limits to fishermen.

American regulators hope it's some insurance against future declines, giving lobster more time to mature and reproduce. It all has a lot of people in the industry wondering what's coming next as the era of peak lobster fades. But fishermen are too overinvested to pull back on the reins. Richard says it's become a volatile time in an industry where not that long ago it seemed like the good times would never stop. He talks to buyers and fishermen across the industry, and he knows how the sector has been devastated farther down the coast, especially in southern New England where the lobster fishery is a shadow of its former self.

"There's no university course in the world that can teach you all this crap," Richard says. "It's changing at such a rapid pace because, you know, what used to be a community industry became a provincial industry and then it became a Canadian industry and now it's become a global industry."

The lobster rush of the past two decades also brought a flood of new players into the industry, where boom-and-bust cycles have long been a given. An American company, Lobsterboys, bought a warehouse on the Dipper Harbour waterfront in 2022 and immediately began striking deals with fishermen and ruffling feathers with its brash style in a community where family connections go back generations. The company's co-owner, Justin Maderia, used shiny golden lobsters and the slogan "luxury delivered" in his marketing efforts, and stood out among the plaid-shirted fishermen with his slicked-back hair, flashy necklaces, and sunglasses, a style that could be described as Peggys Cove meets Jersey Shore. The Maderia brothers came from Stonington, Connecticut, which at one time was in the heart of America's lobster region. But after lobster stocks collapsed in local waters, buyers there had to look farther north to find product. Almost as soon as Lobsterboys arrived, however, the company was dogged by complaints of unpaid bills, and it had to restructure its U.S. affiliate amid legal battles south of the border. By the end of December 2023, the company's doors were closed in Dipper Harbour, and Maderia was complaining on Twitter that feminism was "ruining civilization." As part of a bankruptcy process approved in April 2024, Lobsterboys sold the Dipper Harbour facility to help pay down their debts.

Down the road in Bay Shore, New Brunswick, another seafood packer was bought by a Chinese entrepreneur named Nathan Song, who immediately began shifting focus almost entirely to Asia, selling 90 per cent of the lobster he buys to China.

Richard understands the outside interest in a local fishery no one cared about just a few decades ago, but thinks these newcomers aren't going to stick around in fishing communities once there's no longer money to be made. Lobster has been a lifeblood for Dipper Harbour, bringing jobs and money here, just as it has for so many other small fishing communities on the Atlantic coast. If it goes, he wonders what will be left behind.

"Who's going to be left to mow the cemeteries if we sell it all out?"

3

A Whale for the Killing

NANTUCKET ISLAND, Massachusetts – Whaleships no longer crowd the harbour in Nantucket, and the sparkling white yachts and pretty sailboats outnumber the commercial fishing vessels by about a hundred to one. The shallow water here has become a summer playground for the rich, filled with people who earned their money on the comfort of land.

But two centuries ago, this crescent of sand about thirty miles south of Cape Cod was the whaling capital of the world, and the oil produced by that bloody business made Nantucket one of the wealthiest places in America. It was, if you were a whale and unlucky enough to be passing by, the last place you might see before you were butchered by men in small boats and rendered into oil for lamps or candles.

By 1775, half of the whaling vessels in Massachusetts were based in Nantucket, and the island could produce more than thirty thousand barrels of oil a year. When the island's whalers killed off nearly all the North American right whales that grazed in the island's shoals, they built bigger ships and set off across an ever-expanding territory in search of other, bigger whales. These legendary trips sometimes took as long as three to five years, and reached from the Arctic Circle to the southern Falkland Islands and along the east coasts of North and South America and the western coast of Africa. The whaling industry formed the basis of such a lucrative trading partnership with England that Nantucket Island remained neutral during the American Revolution.

Whaling and the men who stalked the massive creatures across the ocean inspired legendary tales from Nantucket, most notably Herman Melville's 1851 classic, *Moby-Dick*. The tragedy that inspired Melville to write his novel was the final voyage of the Nantucket whaler *Essex*, and his two main characters, Captain Ahab and Starbuck, are both from Nantucket Island. "Two thirds of this terraqueous globe," wrote Melville, "are the Nantucketer's. For the sea is his; he owns it, as Emperors own empires."

Whaling was also incredibly dangerous work, so much so that by 1810, nearly a quarter of women over the age of twenty-three in Nantucket had lost their husbands to the sea. Whaling had created an island of widows.

By the early 1890s, commercial whalers had hunted right whales in the Atlantic to the brink of extinction. Whaling is no longer an industry on Nantucket, nor anywhere else in North America, but humans are still the greatest danger to the species. Vessel strikes and entanglement in fishing gear are among the leading causes of North Atlantic right whale mortality.

It's all left Nantucket, once perfectly situated to be the world's whaling capital, at the epicentre of the fight to save the right whales. Right whales, as baleen whales, are like giant seagoing cattle that feed on small marine animals by straining huge volumes of ocean water through their bushy baleen plates, which act as a sieve. The shallow waters around Nantucket are an ideal feeding ground, right on the whales' migratory path along the Atlantic coast. Every year, these giants travel about a thousand miles from their feeding grounds in Atlantic Canada and New England down to the warm waters off Georgia and Florida, where they give birth to and nurse the precious few calves.

This long, migratory "superhighway," as the Reuters news agency calls it, puts them in direct conflict with the commercial fishery and shipping industries that ply North America's east coast, a problem that's been widely reported by media outlets around the globe. Entanglements with fishing gear killed at least nine North Atlantic right whales between 2017 and 2022, making it the second-biggest cause of death behind strikes from boats and ships, according to the U.S.'s National Oceanic and Atmospheric Administration, or NOAA. That's a big number for a species on the brink, given that fewer than 360 North Atlantic right whales remain, including just around seventy breeding females.

The U.S. and Canadian governments have tried imposing new regulations on lobster and crab fisheries in recent years, including the use of fishing zone closures during whale migration periods, weak "breakaway" links in ropes designed to break if a whale swims into them and begins to get entangled, more easily traced colour-coded rope, and increasing the number of traps per buoy line to reduce the number of ropes leading to the surface.

But whales are still getting entangled. In the spring of 2023, four North Atlantic right whales were hurt after getting caught up in fishing rope, according to U.S. government data that tracks injured whales by air and water. That data isn't definitive—it's believed other entanglements aren't recorded at all, because some whales swim far away from human eyes and are never recorded by officials. One whale spotted off North Carolina was trailing a pair of lobster traps that authorities believe came from Nova Scotia, hundreds of miles away.

Today, the commercial lobstermen remaining in Nantucket argue it's the whales—or at least regulations intended to protect the creatures— that are killing *them*, economically speaking.

Jim Sjolund, one of only two lobstermen left on Nantucket, joined the fishery in 2023 after buying and restoring a used thirty-six-foot boat that he sailed back from Maine. He renamed the vessel the *Julie*

Alice, after his mother and grandmother. He'd spent years working with a cod fishing crew in Alaska and was looking for work that would bring him back home. Lobstering was the only job that made sense.

Jim's family has fished around Nantucket for three generations. His grandfather Rolf Sjolund landed on the island in 1929 and worked the productive cod grounds off Georges Bank, a massive underwater plateau fed by the cold Labrador current off the coast of New England. The ocean just seemed to call to him. He'd left his birthplace of Norway at fourteen, sailing with the merchant marine for seven years before settling on the island with his wife. They raised four kids, including a son, Carl Sjolund, who began harvesting scallops in Nantucket Bay at the age of twelve.

Carl's son Jim, a short, sensible, hardworking type in his mid-thirties, entered the lobster fishery at a difficult time. The easy days of lobster fishing close to Nantucket's shores are long gone. To get to his traps, he needs to steam twenty miles due east over the rough Nantucket Shoals, a series of sandbars that have claimed hundreds of ships over the centuries. To make the fourteen-hour return trip, the young lobsterman had to retrofit the *Julie Alice* with larger gas tanks.

Many people his age who stay on Nantucket end up in the tourism trade or working as carpenters, landscapers, or plumbers for the wealthy homeowners who have bought up the island's shoreline. Others have gone into aquaculture—but "watching oysters grow" is too boring, Jim says. He prefers a life at sea, like his grandfather and father, something fewer and fewer Nantucketers are doing. Long gone are the days when a scallop dragger could easily find a crew on Nantucket's cobblestone streets. No one wants to get up well before the sun and put in long days on the water anymore, Jim says.

"Somebody has to go do this work. Somebody has to go and bring this stuff in," he tells me. "But there's so many other jobs in Nantucket people would rather do now, and they pay more. You just don't see young people going fishing anymore. And that's slowly happening in a lot of places."

The pace of change has been dramatic in the past twenty years. When Jim was a boy, bay scalloping, or harvesting shellfish from local waters, was a popular winter fishery that probably employed around fifty boats on Nantucket. Today, only a few dozen men are left doing this—and they're all old.

Meanwhile, the warming waters around Nantucket are almost void of fishing boats. The shallow waters surrounding the island are only sixty-five feet or so deep, and heat up notably in the summer. In place of boats pulling up lobsters, the ocean here is full of seals, a new arrival, which, along with hordes of ravenous cormorants, Jim blames for devouring whatever juvenile lobster are left.

Jim bought his boat before the right whale regulations took effect, and now he's having second thoughts as the new rules around equipment, fishing zone shutdowns, and boycotts squeeze the ability of lobstermen to make a living. On top of that, he's paying a premium for fuel and bait because he lives on an island, and everything he needs to run his boat must come from the mainland.

Regulations requiring break-away ropes mean it takes Jim significantly longer to prepare his gear. The regulations also require a minimum fifteen traps per trawl, a rule intended to reduce the number of lines reaching from the ocean floor up to buoys at the surface. With his smaller boat, he can't take more than eighty traps with him at the best of times, and all the extra and heavier groundline required because of the new trawl rules means he's even more limited in how much gear he can take when he heads out to sea. Plus, all that extra line on his boat's deck dramatically slows down the pace of his work, and makes it hard for him to get around safely while out on the water.

It's estimated that more than 90 per cent of the commercial fishing ropes in the water in New England come from the lobster industry. Although some fishermen in New England and Atlantic Canada are testing ropeless gear, Jim sees an industry-wide transition to on-demand, buoyless traps as a pipe dream—an unreliable, costly, unnecessarily complicated technology. He says the cost alone would force some fishermen to quit.

"In a dream world, it sounds wonderful," he says. "But it doesn't work. It's not as simple as people think."

Environmentalists argue that the industry bears responsibility for killing right whales and can't blame other fisheries or marine activity—they say fishermen must change their ways to protect the whales.

"The notion that they're not part of the problem is unbelievable," says Kristen Monsell, a senior lawyer with the Center for Biological Diversity, a deep-pocketed Arizona-based conservation group that is dedicated to protecting endangered species and wild places. The group has been one of the loudest voices criticizing the delay in protections for right whales. Its co-founder, Kierán Suckling, grew up in Cape Cod, near the heart of Massachusetts's lobster fishing industry. Suckling, an activist *The New Yorker* magazine described as a "trickster, philosopher, publicity hound, master strategist, and unapologetic pain in the ass," has become an expert at using litigation and the media to turn the Center for Biological Diversity into one of the most successful environmental groups in the U.S.

To many lobstermen, especially those who see themselves as stewards of the sea, environmentalists like Suckling are undoubtedly a pain in the ass. But they're also effective. Activists have used the power of boycotts to convince consumers that the lobster industry is killing right whales.

"I don't think anyone wants their lobster with a side of whale," Kristen says. "There's now more people studying right whales than there are right whales left."

Some see it as an either-or proposition, arguing that the measures to protect the whales go too far and are doing irreparable harm to a fishery under intense financial and climate stress.

"They want to save those whales and do away with the lobster industry," says Sid Look, owner of Look Lobster dealership in Jonesport, Maine.

The vast majority of New England's fishery, about 80 per cent, is dependent on lobster. Other inshore fisheries, including herring, shrimp, and groundfish like cod and haddock, have all but disappeared, shut down by moratoriums and permanent closures as those stocks collapsed below sustainable levels. Lobster catches peaked in Maine in 2016. There is no obvious alternative to lobster.

"If there's no lobster in Maine, then Maine is done, as far as the commercial fishery," Sid says. "There's nothing left for them to catch. That's the damn problem. They've taken everything else away from it."

Fishing zone closures, though supported by some scientists and environmentalists, are controversial. Lobstermen in Massachusetts, which has the second-largest lobster fleet in the U.S. after Maine, are now kept off coastal waters from February through May, while right whales typically migrate, meaning some fishermen can't set their traps until June. From a purely fishing point of view, that's a bad time to begin, because by that point in the season lobsters have begun to moult, or shed their shells.

"It's terrible timing because they've gone into the shed," says Beth Casoni, the plainspoken former cop who is now the executive director of the sixty-plus-year-old Massachusetts Lobstermen's Association in Scituate, a seaside town south of Boston where fishing has been a part of the economy since British colonists arrived in the early 1600s.

Adult lobsters typically shed their hard outer shell, or exoskeleton, once a year or every few years, part of the growing process that leaves them particularly, but temporarily, vulnerable to predators. Fishermen don't want to catch moulting lobsters, since those with hard shells are best for eating, but the right whale restrictions in the state mean the delayed fishing season coincides with the early-summer moulting period.

Frustrated at being kept off the water during prime lobster season, some Massachusetts fishermen have turned to the courts to fight for their right to harvest. The Lobstermen's Association also set up a legal defence fund to push back against restrictions, paid for by fishermen, but it's only able to contribute a few hundred thousand dollars to lawyers' fees. Lobstermen simply can't compete with environmental organizations that have the ability to raise tens of millions in donations, Beth says.

"If you put a picture of a lobsterman and his family on their front lawn with a foreclosure sign and something that says 'Help,' people will say, 'Get another job,'" Beth says. "But if you put a picture of a whale wrapped in rope and a sign that says 'Save Me,' people will throw millions of dollars at you."

The lobster fishery in Massachusetts is a fraction the size of Maine's, but it's still a valuable piece of the coastal economy. In 2021, lobster landings were worth $125 million to the state's fishermen—more than $50 million beyond their 2015 value.

The Lobstermen's Association, with its two paid staff and handful of volunteers, is trying to counter the message being promoted by some activists that buying lobster is bad for whales. But with its limited budget, it's hard to make that argument heard away from seaside communities in New England.

"We're doing our best," Beth says, "but some people just don't want to hear it. They think lobstermen are the antichrist. It's sad. Financially, we can't compete with their war chest."

Lobstermen in New England and Canada complain that environmentalists have convinced the public that ropeless, or "on-demand," gear is the answer to saving the whales. Beth disparages the idea that what she calls "Gucci" traps, costing $3,000 or $4,000 apiece compared with $80 or $100 for a conventional wood and wire trap connected to a buoy with polypropylene rope, are a feasible option for most fishermen. Considering that lobstermen in the U.S. can put out up to eight hundred traps, the price to protect whales quickly becomes prohibitive.

Ropeless gear, which is not yet widely used commercially, works by suspending ropes in the water only briefly when a lobster boat arrives to retrieve its traps. One of the most promising models uses GPS tracking to locate the traps on the ocean floor, then remotely triggers a rope and buoy to be released to the surface so the string of traps can be hauled up.

Fishermen protest that this technology is not only unreliable and unaffordable but puts their traps at risk of scallop or other trawlers who drag the ocean floor and, because there are no buoys to mark their location, don't know there are lobster traps in the area. Beth thinks a better, cheaper solution is break-away ropes. Most fishermen long ago switched from the old manila or hemp lines to stronger polypropylene rope, which is good for saving traps in rough weather, lasts longer in extreme ocean conditions, and reduces lost gear, but may be contributing to entanglements.

But even talking about a mandated switch to ropeless gear, which could be just a few years away, gets Beth's blood boiling. If the cost doesn't come down, some fear it could make lobstering financially unfeasible for a lot of independent harvesters. Not to mention unaffordable for consumers.

"Nobody is putting their hand in their pocket to buy gear that costs that much," says Beth. "It's a long way off, but I can't get my blood pressure up right now."

Nevertheless, both the American and Canadian governments are promoting ropeless fishing gear as a possible long-term solution to whale entanglements. In 2022 NOAA published a strategy document that outlines a path for the adoption of ropeless technology in coming years, and Canada's Fisheries Department has called ropeless gear "the only way you eliminate the vertical line entanglement risk."

Right whale regulations are coming at a time when U.S. national and state governments are also prioritizing offshore wind development, another challenge for fishermen trying to protect their turf. Combined with the very real impacts of climate change, the industry is facing what

Rick Wahle, a professor with the University of Maine's School of Marine Sciences and one of the world's leading experts on American lobster, calls "an existential threat on both sides."

Some attitudes may need to change when it comes to doing things differently in the name of sustainability. But among fishermen, there's still some reluctance to try the new ropeless lobster gear.

"No one wants to be at the front of that because of the repercussions," explains Sam Belknap of the Island Institute's Center for Marine Economy, "because it's seen as anti lobster industry. There have been threats, toward gear, to your boat, your life, your home or your family, if you're perceived in some instances to be part of the problem. If you are testing this on-demand gear, that's a faux pas. In many cases, you're perceived to be part of the problem that you're seeking to figure out if that's going to work.

"But the flip side of that is if we're not investing the time and effort to figure out how it works, where it works, and how it can be improved, before we're forced to use it, then we're going to be in an even worse position when some of the regulations hit in five years, irrespective of whatever climate change keeps piling on top of us and the ecology and the population of the lobster alongside that."

Three days before Christmas in 2022, the Democrat-controlled U.S. Senate passed a massive omnibus funding package that put a six-year delay on stricter measures to protect the right whales. It bought lobstermen a bit more time in their campaign for more balance in the protection measures. Environmentalists, meanwhile, feared the delay would push the whales to the brink, pointing out that, according to the country's federal fisheries agency, U.S. fisheries entangle more than 15 per cent of the dwindling right whale population each year.

Jim Sjolund says the regulations are changing so quickly, the industry can barely adapt, and there's hardly enough time to gauge whether those changes are effective.

"They've been changing the rules so fast they don't even know what works," he says.

He thinks some activists won't be happy until nearly all commercial fishing activity is banned. Like a lot of lobstermen, he gets heated as he talks about the changes that may be coming.

"What's their end goal? They want to get rid of everybody to save the ocean? I don't get it. That's a real knee-jerk reaction. There's been lobstering, long-lining, and gillnetting for hundreds of years," he says. "I think it's ridiculous."

4

The Bleakness

STONINGTON, Connecticut – The *Lady Lynn* is not going to sea today. Michael Grimshaw, the boat's sixty-six-year-old owner, has his head stuck inside its engine room with a ratchet in his hand, launching a streak of curses at the big diesel motor that belches smoke and spews seawater every time he turns it on.

The forty-one-foot fibreglass lobster boat has been gutted and rebuilt so many times since it started working the waters of Long Island Sound in 1981 that the only original part left is the cracked compass glued to the dashboard. Once again, the *Lady Lynn* needs more repairs. Not that it really matters anymore.

"This fucken' thing," Michael says in his heavy New England accent, and trails off.

He bought the vessel when he was in his early twenties, the year his mother died. In flat seas, it rides like a bucking bronco. But in heavy seas, the boat's design helps it cut through the waves clean and steady. The worse the weather, the more stable the boat seems to get.

The *Lady Lynn* has been grounded, nearly sunk, swamped, and seized by conservation officers when Michael was caught harvesting scallops out of season on the border with Delaware. They found a bag of marijuana behind the windshield, he laughs. Docked at the government wharf in

Stonington, tucked among the idled scallop boats, it looks like it probably should have been retired years ago. But there's no point in replacing it now, and no money to do that anyway. Not since the lobster disappeared from Long Island Sound, the long, shallow body of water between the Connecticut shoreline and Long Island in New York to the south.

Michael is the last full-time lobsterman still working these waters. He remembers the good years, when there were so many lobster they needed to use a forklift to carry their catch out of the boat. Today, he's lucky if they can fill a few plastic bins.

The rapid decline of lobster in Connecticut seemed to hit fishing communities here like a hurricane that suddenly veered off track and turned inland. But there were plenty of warning signs that went ignored, even as fishermen reported record catches around Long Island Sound in the years leading up to the collapse in the fall of 1999, according to a study by University of Connecticut researchers. A year before the collapse, many lobstermen in the sound's most western reaches found their traps were coming up empty. In 1997, larger than normal amounts of lobster in the area around Greenwich and Stamford were already dead when fishermen hauled them out of the water. One dealer reported that lobsters from Long Island Sound were dying in his holding tanks and those of his customers.

By August of 1999, fishermen began reporting "lethargic, moribund, and dead lobsters" to state officials in both Connecticut and New York. That fall, the catch had shrunk nearly 99 per cent from the previous year in the sound's most western waters, and as much as 90 per cent farther east. By the end of the year, the governors of both states were requesting disaster assistance from the federal secretary of commerce, William M. Daley, who declared the lobster fishery in Long Island Sound a "commercial fishery failure due to a resource disaster" and pledged $13.9 million of federal disaster relief funds.

Connecticut's historic lobster fishery has never recovered. This in the state that brought the world the lobster roll—invented at a Milford restaurant named Perry's, where the first documented lobster sandwich, with lobster tossed in butter and served in a grilled hotdog bun, was served up in 1929. For fishermen here, with a proud history of lobstering in a state that was once in the heart of America's lobster fishery, the sudden die-off that happened in 1999 as water temperatures spiked has marked a painful turning point in their entire way of life.

Michael calls it "the Bleakness." A heavyset man in a stained grey hoodie and a beige cap, he's logged more miles at sea than most fishermen in this part of the U.S. And the *Lady Lynn* looks like it's seen every square inch of water here, with its faded peach paint job and sunbleached American and Jolly Roger flags flapping in the wind.

Michael named the boat after his ex-wife, but as any good, superstitious lobsterman will tell you, changing a boat's name is bad luck. So he rides to sea on a boat named after a woman with whom he no longer speaks. He hangs a painted rubber mermaid from the *Lady Lynn*'s mainmast, for good luck. His first wife painted the mermaid's hair black, like hers. When he remarried, his new wife painted it blond.

He laughs when he says this, then immediately starts yelling at a young deckhand who has flipped the wooden engine cover over on its back.

"That means your boat is gonna sink," he explains.

Lobster fishermen in Long Island Sound don't need to tempt fate. A memorial at the far end of the wharf lists the names of hundreds of fishermen lost at sea. Many of them are Michael's relatives. He's had his own close calls, too. Fishing has beaten up his body, ruined his first marriage, given him a stroke, and forced him to get two knee replacements so that now he has to sit on a stool when he drives the *Lady Lynn*.

He fell overboard one February and nearly died from hypothermia. He was in the frigid water for about forty minutes before the fire department could pluck him out. Paramedics had to cut him out of his waterlogged clothes so they could begin to get his body temperature up.

"They were new jeans, too," he says.

Michael loves being a fisherman. But he realizes it no longer makes any sense to keep going, now that the lobster are pretty much gone.

"Anybody who wants to get into this business now has to have a screw loose," he says.

He's only doing this to keep his son, Rodney, away from some of the problems he was getting into on land as a teenager. Putting him to work lobstering was the best way Michael could think of to get him on a different path and try to instill some work ethic in him, he says.

"I don't know what the future is. I mean, right now I'm just trying to keep my kid out of trouble," Michael says. "They want to hang his ass out to dry. I mean, years ago we went around as kids smashing mailboxes with a sledgehammer. And you know that's a federal offence, right?"

Lobstering is also hard on those who stay behind and remain onshore. Fishermen regularly get up around three in the morning and don't return home until evening. When the fishing was good, Michael would be working seven days a week. His last marriage ended because his wife got tired of him always being out in his boat. On Sundays, when he wouldn't go fishing, he'd be down at the wharf tending to his traps and gear.

But when the boom was on in Connecticut, nothing beat the rush of lobstering. Michael loved being out on the water, watching the hoist pull up line after line of traps full of squirming lobsters.

"That was better than any casino for me," he says. "It was like pulling on the slot machine lever. You never know what you're gonna get in the trap, right? I used to get my charge from that. Especially when it was at its peak, every pot coming up was beautiful. It was just beautiful." In his best years, he'd haul in as much as $700,000 in lobster. Today, he's lucky if he can cover the cost of his diesel.

He says, "I think back on those days like, 'Did it really happen?' I saw the best years out here. That was until the Bleakness."

———

Long Island Sound is the southernmost inshore range for American lobster, and the industry's collapse here serves as a warning for fisheries farther north as ocean temperatures rise and climate change alters the complex marine ecosystem fishermen depend on. According to Nancy Balcom, a University of Connecticut biologist, the water temperature here increased as much as a full degree Celsius in the past three decades—enough to significantly impact creatures as sensitive to water temperatures as lobster.

For years, warming waters seemed to be fuelling a boom in lobster populations. Then, suddenly, a tipping point was reached. Between 1997 and 2001, new diseases emerged in Long Island Sound. Perhaps the worst outbreak occurred in 1999, when storms drove warm surface waters deep into lobster habitats at the western end of the sound, where it fed algal blooms that severely impaired the lobsters' immune system, and many succumbed to infection. J.B. MacKinnon, writing in *The New Yorker*, described how this phenomenon allowed a pathogen called epizootic shell disease to wreak havoc on shellfish populations.

"The warmer water fuelled algal blooms that robbed lobsters of oxygen, impairing their immune systems so severely that many succumbed to infection from a seafloor-dwelling amoeba that is usually not parasitic. The most lethal illness in the rest of sound was a newly described pathogen called epizootic shell disease, which continues to be a problem. All told, Long Island Sound lobster landings have declined by ninety to ninety-five per cent," he wrote.

As the lobster disappeared, many fishing families packed it in and headed elsewhere for work. Most sold their boats to fishermen farther up the coast in Massachusetts or Maine. Those who stuck around and kept their boats all have day jobs to pay their bills. Michael says that nowadays, it's a job just to find someone to sell you fresh bait for your traps.

Just down the coast in Noank, Connecticut, a scenic town of nineteenth-century wooden homes and white picket fences crowded onto a peninsula at the mouth of the Mystic River, the only lobster on the waterfront are those served up in forty-dollar sandwiches to the crowds at Abbott's Lobster in the Rough and Ford's Lobsters. From this former fishing village, fishermen used to scatter out along the coastline, filling up their Noank smacks—boats with circulating seawater tanks in their hold—to the brim and carrying their catch to New York and Boston. So many lobsters were being unloaded at the waterfront that Abbott's installed a conveyor belt so the boats could dump their catch straight to the wholesalers onshore.

Today, those working boats have nearly all been replaced by gleaming white yachts and expensive speedboats, driven by affluent new homeowners from New York City who've turned the town into a retirement community for the moneyed. Michael remembers when these former lobster ports were working villages where people had to earn a living on the water, not line up for tables to dine near it.

"It rips my ass," he says, "that all these people come into town and they buy all the properties up and put them in such a high price bracket that the average person that would have been a fisherman can't live in this town anymore. You know, there's no more grocery stores in town. There's no more drugstore here for us in the village. And that was the end of that."

Noank fishermen used to haul in half of Connecticut's entire lobster harvest. In the nineteenth century, the town was the centre of the state's lobster fishery, and played a critical role in the commercial fleets that carried lobster from Boston to New York.

In 1898, a sudden die-off occurred here, but there were warning signs about overfishing long before that. Federal fisheries wardens had already reported that the lobsters being caught along the Connecticut shore were getting smaller and smaller—so much so that five- and three-pound lobsters were increasingly rare, replaced by lobsters two pounds or less.

Newspaper accounts from the era describe catches that were about a tenth the size of what lobstermen were hauling in thirty years earlier.

In 1896, a devastating heat wave set the stage for even sharper declines, turning the live-well smacks that had made Noank famous into death traps for lobsters. The inshore fishery collapsed up and down the southern New England coast, forcing state officials to pour money into hatcheries in a desperate attempt to restock the fishery. It would take six decades for lobster to return to Long Island Sound, at least in any level that was commercially viable. Today in Connecticut, even the local lobster wholesaler parked at the Stonington wharf, with a cartoonish shellfish on its trucks, buys its lobster from Canada.

Still, the last lobsterman in Long Island Sound keeps heading out onto the water.

"I don't think anybody's earning a living anymore," Michael says, wiping his brow. "They're all in the same boat. Most of the people that were the players, if you will, the people who were the high-liners, are either dead or too goddamn old and retired, and there's no way they can do it. So yeah. It sucks." He looks over at his son, and then looks out to sea. Then he sighs. "He'll never see what I saw," he says.

5

Grandfather Lobsters

BAY OF FUNDY, New Brunswick and Nova Scotia – Once in a while, the crew of the *Small Fortune's* pull up a barnacle-covered specimen so big and so old that it looks like it carries a lifetime of secrets from the deeps. The fishermen will take turns admiring it, sometimes posing for pictures, before its massive claws are closed with multiple rubber bands and it's added to the rest of the catch. Chances are it will be on a dinner plate in China within a few days.

"When we find a great big one, one that's really ugly, we call them dinosaurs," shouts first mate Tom Duke over the roar of the wind and the groaning of the boat's motorized hauler. "They're like grandfather lobsters. They beat all the odds to make it."

Thanks to an enzyme called telomerase, lobsters never stop growing, whereas most animals cease growth after they reach sexual maturity. While they never *stop* growing, that growth does slow over time. Eventually, it takes too much energy to keep moulting and growing, and lobsters will die. A nine-pound lobster like one Tom pulled from the Bay of Fundy could be over fifty years old, although it's difficult to precisely determine the age of lobsters. Accounts from fishermen in the nineteenth century in North America regularly reported catching lobsters as much as three and four feet long. Some truly gigantic specimens are still found, although they're caught with less and less frequency. The largest on record, a forty-four-pound behemoth, was caught in Nova Scotia in 1977 and was estimated to be over a century old. Most of the lobsters that consumers see in grocery stores and restaurants are far younger, however, typically under eight years old.

To survive to adulthood, a lobster must run the gamut of obstacles in an unforgiving environment where everything seems stacked against them. For every fifty thousand eggs a female lobster releases, only about two juveniles are expected to make it to legal harvesting size—a process that can take between five to seven years. That means a one-pound lobster must successfully spawn at least four times to guarantee at least one of her eggs becomes a full-grown animal. The larger the lobster, the more her egg production increases, which is why the biggest lobsters matter so much for the reproductive health of the species. A nine-pound female may carry more than a hundred thousand eggs, whereas a one-pound female usually carries around eight thousand.

Lobster life begins as it does for many creatures: as an egg, carried around inside a female for nine to twelve months. The eggs can spend up to another year outside the mother's body, attached to the underside of her tail, where a glue-like substance holds them to her feathery pleopods, or swimming legs also known as swimmerets. In this stage, they look like tiny blackberries, which is why some fishermen refer to egg-bearing females as berried. In Canada and most U.S. states, fishermen who are caught with a berried lobster face a $1,000 fine. Lobster fishing is unique in this way; no other fishery throws back a part of its catch to ensure the next generation's survival.

When the eggs finally hatch, the female lobster fans them off her body by flapping her swimmerets. This typically happens when the ocean is at its warmest, in the summer months. The lobster larvae float up to the surface and spend months getting pushed around by the current. They're incredibly vulnerable at this stage, and most will be gobbled up by fish while they bob around helplessly. Baby lobsters will moult up to twenty-five times during this time.

Even as they grow larger and better able to protect themselves, lobster remain skittish, perpetually afraid of being eaten. They hunt in the darkness, waiting for the sun to set before they begin searching for food. They aren't picky eaters, hunting large amounts of mussels, crabs, sea urchins, marine worms called polychaetes, and small sea stars. Once they reach adulthood, with hard shells and sharp claws to protect them, they must still moult as they keep growing, which again leaves them vulnerable to predators. Males will shed their shells on average once a year, females once every two years. A lobster caught in spring is more prized because it has a hard shell full of meat compared with one caught in August, when its shell is still new and soft and it hasn't yet filled out.

Lobster used to have a lowly place among scientists who study ocean creatures, dismissed as giant cockroaches of the sea. But by the early 1980s, studies were beginning to show just how complex and sophisticated they really are. Those studies revealed that lobster have a taste system that's a million times more sensitive than a human's. Using tiny hair-like receptors, a lobster can detect chemical compounds over extremely long distances, helping it find food many miles away underwater. This is why those foul-smelling bait bags that fishermen use to lure lobster into their traps are so effective. Because of that highly developed sense of taste and smell, lobstermen prefer bait that's begun to rot, so as to attract lobster from across a wider range of the ocean floor. The smellier, the better, or so the thinking goes.

"Lobster aren't like you and I, going into the grocery store and checking the produce to make sure it's fresh. They're scavengers, they'll eat anything," says Brad, the captain of the *Small Fortune's*.

Lobster operate on ancient seasonal instincts that tell them when to migrate, when to gorge themselves, when to moult, and when to mate. For centuries, very little was known about the secret lives of lobsters.

Over time, marine biologists began to build an understanding of the animals' mating habits, their ecology, migration, and other behaviour patterns. To do this work, they often had to enter the lobsters' under-water world, using submersibles, snorkel equipment, scuba gear, and trawl surveys, in which research vessels drag a large net across the ocean floor and examine what they pull up. Generally speaking, most lobster are migratory, moving south in the fall and north in the spring. One surprising new revelation is that lobsters can use their sensitive hair-like receptors to find their way across extremely long distances of ocean floor, fighting tides, powerful sea currents, and other obstacles, to return to their homes, where they can hide out for weeks or months at a time. One lobster tagged in a study off the Continental Shelf was recovered at Port Jefferson on Long Island, New York, a distance of more than 225 miles.

Bayard Webster, a science editor for *The New York Times*, wrote in 1982, "The simple fact has been determined, for instance, that this noc-turnal, solitary, aggressive-looking invertebrate, who may live to the age of 40 or more, can home like a pigeon. It can always manage to find its way back to its own burrow, although it can barely see."

The lobster's strong sense of smell also plays a key role in its mating habits. Females release their pheromones in the water as a sexual attrac-tant, typically in front of the entrance of a male's den. The male takes this as a sign to let her into his den, which can be a crevice in a rock or a shelter he built by piling up mud and pebbles, where she'll moult and then mate within half an hour of losing her shell. After mating, the couple will live together for a week to ten days, so the female will be protected while her shell is hardening. Males have been observed block-ing the entrance of their den with their big claws to protect the female until her new shell was hardened enough for her to venture out.

Mating is only the first step in a two-year process for the female to finally produce baby lobster, explains Heather Koopman, a marine biol-ogist at the University of North Carolina Wilmington who works at a research station on Grand Manan Island in New Brunswick during lobster season. That's why protecting adult females, even those that don't appear to be carrying eggs, is so vital for the survival of the species.

"Once she's shed her shell, that's the only time that a male can mate with her," Heather says. "Then she stores the sperm in a little sac, and for a whole year after that, her eggs develop internally. So if you look at the lobster, you can't tell anything's happening. But her eggs are developing internally and finally, the next summer, she will extrude her eggs. They come out of the internal ovary and she deposits them on the underside of her tail, and that's when she fertilizes them. And she carries these eggs around underneath her tail for another whole year, while aerating and protecting them."

As scientists began to unravel the mysteries of lobster, they also got better at measuring the overall health of the population. Those studies have made it clear that the abundance of lobster is not permanent. Like all heavily fished, wild-caught commercial seafood, lobster are sensitive to overexploitation. Although fishermen are catching more lobster today than they were a generation ago, a fact that's sometimes pointed to as proof of a healthy resource, the number of traps and the general efficiency of the fishing industry have also increased dramatically.

One troubling consequence of decades of intensive fishing is that jumbo lobster are increasingly rare in the North American fishery. That's not just a problem for egg production. As fishermen take away the largest lobsters from the species, they're also removing the best warriors, allowing crabs and other rival animals to dislodge a lobster population from an area.

Back on board the *Small Fortune's*, Howie wonders aloud why fishermen were ever allowed to keep lobsters that big in the first place, given their important place in lobster hierarchy. In Canada, there is no maximum size limit, whereas there is in most jurisdictions in the U.S.

"They're your best breeders," he says. "And now we're targeting the best breeders, just because they want them in China."

It's an issue on both sides of the Atlantic. The largest European lobsters, the American lobster's blue-shelled cousin, are also disappearing because they've been overfished for many decades by fishermen seeking them out for export markets.

Julie Hill, sales manager for a Welsh seafood exporter called the Lobster Pot, explains, "In China, they want the biggest sizes, because they use them as gifts. It's a presentation thing. But it's a problem for us, because we don't have a lot of the big ones left."

The arrival of countries like China and Korea as major consumers of North American lobster, and as markets where size matters more than anywhere else, has put a premium on the largest lobsters and complicated conservation efforts. Another fisherman who spent more than fifty years working out of Dipper Harbour, Greg Thompson, thinks that catering to the demands of Chinese buyers is putting more pressure on the resource than it can bear.

"The negative side is we have overexploited fishing in this area, by building bigger traps to catch big lobsters because this is what the Chinese market wanted," he says. "So we've overexploited, in my personal opinion, these larger lobsters and they are mainly our broodstock. And you may note that in the last four or five years, the lobster catch is declining each year but also the catch of large lobsters is declining as well."

Some Chinese buyers will buy only lobsters above two pounds, a fundamental difference from the American and European markets, which have historically been more interested in smaller lobsters. That's partly because of the way lobster is consumed in China—often at opulent banquets, to show off at weddings, or chopped up and fried in restaurants where presentation is as important as taste. The Chinese market also altered the traditional calendar that lobstermen worked around, shifting their focus from American holidays like Mother's Day and the Fourth of July to Chinese New Year, now one of the most important times of year when it comes to demand.

"We have to be thinking in global terms, you know, as a fishery," says Greg, the fourth generation in his family to fish out of Dipper Harbour. "We're exploiting this particular market, but we're doing it at the expense of our future. And I guess it's sort of like the national debt that way. We're expecting our grandchildren to deal with what we've spent today. And I think it's the same in the lobster industry. We're going to find out that our children and grandchildren are going to have a much different fishery than the one we've enjoyed."

As the market for larger live lobsters, which used to be sold at lower price to canneries, has grown, technology has allowed fishermen to venture into areas they previously couldn't fish. A generation ago, fishing gear limited fishermen to working closer to the shoreline because their buoys and ropes couldn't handle the stronger tides and bigger depths farther offshore.

"In my lifetime, the technology has increased the catch tremendously," Greg says. "When I first started, fishermen always fished within a mile of shore. Then it wasn't too long until we got up to three miles from shore. But for a while, we couldn't really go much farther because the tides would take our buoys under, the buoys would shrivel up, you'd lose your gear and you didn't really catch much."

Greg, who retired in 2022 when he found his body couldn't keep up with the work in the way he wanted it to, watched the evolution happen right in front of him, on the wharf and on the boats that crisscross the Bay of Fundy to get to their secret lobster grounds. Over time, wooden traps were replaced by more durable wire traps, stronger, synthetic polypropylene rope replaced natural-fibre rope, and bigger boats were built that opened up whole new patches of ocean that had previously been off-limits to fishermen. Eventually, there was no place left in the fishing district where lobsters could hide.

"And all of a sudden we were able to exploit this, probably at least 75 per cent of the district that hadn't been fished before. And so it looked like 'Wow, what a tremendous stock we have, lobsters are booming.'

But it appears as we look now, that, like the inshore, they will not handle the fishing pressure. So we have to seriously look at how we're going to preserve this lobster stock. While we still have time."

There are other warning signs that the lobster population is in trouble. A measure that biologists call the "biomass"—the overall population health of a species in a specific area—is in decline in the southern edge of Canada's lobster fishery, particularly the Bay of Fundy, and in more severe decline in more southerly American waters. Researchers know this because they spend their summers diving and surveilling the ocean floor in traditional lobster grounds, counting the number of baby lobsters they see at the end of black plastic tubes.

Female lobsters are also reaching sexual maturity at a younger age. What's more, not only are larger, super-productive females becoming increasingly rare, but biologists are also concerned about the growing numbers of females hauled up from the bottom with fewer eggs on their tails. It's an important biological indicator of a species going through a dramatic change, according to Rémy Rochette, who runs a lobster research lab for the University of New Brunswick in Saint John and is another of the world's top lobster experts.

Scientists have spent well over a century studying lobster, getting a better understanding of their mysterious life cycle, observing them in the wild and in laboratories. But there remain challenges around precisely predicting how changing oceans and fishing patterns will affect lobster.

"We still don't really understand how these systems work," Rémy says. "We're not that good at forecasting."

If it weren't for the growth in the catch in the fishery's northern regions, the lobster industry would overall be in decline. And a fisherman in Maine or southwestern Nova Scotia probably doesn't care that the fishing is good in the Gulf of St. Lawrence or northern New Brunswick. All they know is their catches are declining and their entire

livelihood is at risk. Scientists can make estimates on lobster population, but it's still just an educated guess. All fishermen can see is the
result in their traps.

"It's incredibly difficult to actually know how many of these darn
things are down there," Rémy says.

But despite the many warning signs, many fishermen are hesitant to
embrace change. Greg, the retired fisherman, suggests part of the problem is the long-held belief in fishing communities that harvests are
cyclical, that downturns are followed by a return of good harvests. He
says that kind of thinking doesn't factor in the harmful effect of modern fishing methods and technology—unseen in previous generations—
on lobster stocks. And too many fishermen still think of themselves as
independent operators, fishing out of their local harbour, far removed
from the global market. They don't think about the collective influence
of their industry on the resource they depend so much on, he says.

"They're still thinking, 'This is what my grandfather told me, this
is what my father told me, this is what it is.' But they're not really considering what they have done in their generation that has affected
things," Greg says. "I've been an advocate all my life of more conservation. But you know, the fishermen will say, 'Well, my father and my
grandfather always said there'd be ups and downs, and you don't adjust
your fishery just because there's a downturn, because they lived through
it. And then another year later, the lobsters were there.' So yes, but we're
doing it differently. The old rules don't necessarily produce the same
results anymore. But it's a tough sell."

Greg knows all about the old rules. When he began fishing, in the
early 1970s, there was a man in Dipper Harbour who still rowed out to
his traps and hauled them by hand, much in the way fishermen did in
the nineteenth century. Greg never planned to make a career on the
water. He started fishing while in university as a way to pay for his
schooling. But after his father suffered a heart attack on the water and
couldn't do the work anymore, he offered his son his boat. His father

died while Greg was still in school, and it seemed inevitable he would continue the family business. Instead of it being a temporary source of income, though, he kept at it for more than fifty years, until his body told him it was time to quit.

Fishermen don't need to wait for regulations to change to begin taking steps to protect lobster stocks, he says. Decades before he began fishing in Dipper Harbour, fishermen there collectively agreed to stop catching undersized lobster, called shorts.

Greg remembers, "They just decided, you know, we're only taking next year's crop, and we're giving it away here because some people sold them cheap. It's just very poor business. And so they decided to stop, and most of the fishermen in the community agreed with them. So the short lobster business in Dipper Harbour dropped back considerably, while it continued in other communities.

"So there's more ways to be conservation-minded than simply obeying the laws. You can go a little beyond that, if you, you know, are serious about the business."

6

Heat Waves

NEW YORK CITY, New York – In the summer of 1896, a heat wave settled over New York City that turned the crowded tenements on the Lower East Side into dangerous pressure cookers. For hundreds of people, living without running water or air conditioning and trying to escape into the streets and onto rooftops and fire escapes, it was a death trap.

The ten-day heat wave that killed nearly fifteen hundred people across the city brought relentless temperatures above 32 Celsius, with 90 per cent humidity and no wind to offer any relief. It also helped relaunch the political career of a little-known police commissioner named Theodore Roosevelt, who became American president in 1901 and championed efforts to give away ice to the poor and open up city parks at night so people could escape the heat. Had his measures been adopted sooner, it's widely believed they could have saved many lives.

New York City in the late nineteenth century was a city of extreme contrasts. While millions of immigrants were crowded into desperate and unsafe living conditions, the era also saw the arrival of "lobster palaces," large, opulent dining establishments that catered to the wealthy, the well-to-do, and elite international travellers. Many of these restaurants opened up in the city's theatre district, and became famous for their exorbitant prices and lavish interiors, with menus that centred around increasingly expensive lobster dishes.

Yet even here, inside these dens of opulence, the ultrarich couldn't escape the effects of a massive warming trend that was directly affecting their menus. In the late 1890s, the same weather phenomenon that hit many parts of North America was also affecting the waters off New York City and New England. There, out of sight from most city dwellers, a similar wave of death was happening that was devastating sea life, most notably lobster. Fisheries managers around the Atlantic coast began noting the dramatic disappearance of cold-water species such as bay scallops, lobsters, and quahogs, while at the same time, soft-shell clams, oysters, and blue crabs were surging to levels not previously recorded. When fishermen in Rhode Island began catching tarpon, a species of fish normally native to waters south of Virginia, they became alarmed.

Fisheries officials in Connecticut, not understanding the influence of water temperature on shellfish, blamed the collapse of lobster stocks between 1896 and 1898 on advances in technology that they believed led to overfishing.

One report from the state's fisheries and game department put it this way: "There has been gradual decline of the industry not only in Connecticut but also in New York and all of the New England States. The beginning of it may be traced to the period when gasoline engines were substituted for sails as propelling power. The power which is used for propelling the boat may also be used for lifting the pots, making it possible for one man to handle a great many more pots than in the early days of the sailboat."

Heat waves have often signalled larger changes in the natural world that scientists are only now beginning to understand. The lobster collapse of the late 1890s was massive, reaching as far north as Newfoundland, where catches also declined suddenly and dramatically. It took years for the fishery to rebuild, and some jurisdictions, including Rhode

Island, temporarily banned lobster fishing. The lobster collapse also led states and provinces to invest heavily in lobster hatcheries, hoping they could restock the wild population.

When the southern New England lobster fishery failed again a century later, it was during another period of high heat. The difference this time is that fisheries managers now understand just how sensitive many marine species are to temperature changes in their habitat, affecting everything from reproduction and growth to predation and disease.

Researchers with the U.S.'s National Marine Fisheries Service estimate "more than two-thirds of marine species in the waters off the Northeast United States coastline are fleeing their traditional habitats in search of their preferred water temperatures," according to a 2018 report by the Center for American Progress, an independent, nonpartisan policy institute. These profound shifts in fish and shellfish distribution can cause further ecological disruptions, the centre reported, by removing food from traditional predators, introducing new predators to the habitat and changing the timing of seasonal migrations. For fishing communities, that can mean economic disruptions as shellfish and fish stocks either decline or migrate away from traditional fishing grounds. The stakes for fishing industries are obviously high. In the U.S. alone, commercial and recreational fisheries supported 1.6 million jobs in 2015 and generated an estimated $62 billion in incomes. Globally, scientists estimate that the fishing industry will lose approximately $10 billion in annual revenue by 2050 as a direct consequence of the expected ocean temperature increases.

Increases in water temperature force lobster to "move" north, in favour of cooler waters—a trend expected to continue, according to projections. As their traditional habitat becomes increasingly inhospitable, lobster are faced with multiple stressors, and the combination of increased warming, toxic algal blooms, and coastal acidification affects the way they breathe and swim. Scientists say that what's happening should be imagined not as a giant underwater northward migration, but rather as changing conditions that allow more juveniles to flourish in more northerly colder water, while fewer are reaching adulthood in warmer waters.

"It's not like adult lobsters are picking up their tent stakes and starting to march northward," says Rick Wahle, the University of Maine lobster scientist. "But on balance it comes down to birth rates and death rates, that there's net losses happening in southern New England and net gains happening in the north. It's a really interesting story, because starting in southern New England, we've already been seeing this since the late 1990s."

There is a lot of evidence that warming waters alter lobster migration, too. But because it takes so long for lobsters to reach market size, typically around seven years, there's a long delay in between problems scientists see with baby lobsters and when problems start showing up in the commercial fishery.

"The signals you might see in female movement patterns or reproduction aren't going to have an impact on the fishery for another seven years," Heather, the marine biologist, says. "A bad year in 2003 isn't going to show up until 2010, or 2023 isn't going to show up until 2030. So everything's time delayed, which makes it difficult to draw a lot of conclusions."

Heather runs a monitoring program that works with fishermen to track the movements of egg-bearing lobster. It's based on a simple approach: When a pregnant female is caught by a fisherman, it's tagged with a zip tie that shows an ID number and Heather's phone number, then it's released back into the ocean. When a lobsterman catches the tagged lobster again, they can show the scientist where they found it by texting her a photo of the animal next to their boat's plotter, which displays the exact longitude and latitudes.

Her research has shown that a lobster caught off Grand Manan can easily make its way across the Bay of Fundy to Nova Scotia in a month or two. Many go hundreds of miles farther. These tagged lobsters give

Heather a rough idea of where the females are migrating over the course of a year, important information for fisheries managers trying to figure out where lobster populations are headed and how to best regulate fishing areas.

In other studies, Heather attaches thermal loggers, which CBC News described as looking like a big yellow wristwatch, and records the water temperature around the lobster every fifteen minutes. This research shows how much lobster eggs require cooler temperatures when they're developing. But once they're laid and stored underneath the female's tail, it appears they need warmer temperatures to hatch at the right time.

"Lobsters are phenomenally temperature sensitive," she says. "They've done laboratory studies where a lobster can detect a difference of 0.1 degree Celsius in its environment. And female lobsters have really strict temperature requirements for successful reproduction."

The warming ocean affects every part of lobster life. Rick and other researchers believe there's a link between recent declines in lobster reaching adulthood in the Gulf of Maine and changes in what's called the pelagic food web, the open-ocean food chain that starts with tiny plant-like organisms called phytoplankton. Scientists have been given significant government grants to better understand this phenomenon and to explore the connection between changes in the ocean's food system and the rapid warming already being documented in the Arctic.

As seawater temperatures rise beyond the ideal thermal range of 12 to 18 degrees Celsius, the lobster's natural self-defences and immune system become increasingly stressed. It's especially a problem for the largest lobsters, who are the most important to the species from a reproduction point of view. And it means that lobsters that are caught in these conditions are less durable and sometimes not healthy enough to survive the long flights to markets in Europe and Asia.

"You start to see their physiological mechanisms shutting down," Rick explains. "As temperatures rise, the oxygen saturation point in water declines, but the physiological demand for oxygen in lobsters and other invertebrates is still going up. So it's a very tenuous situation where they're increasingly constrained by the supply of oxygen. And of course, the larger the lobster, the more its total oxygen demand. And so they're the ones that are most vulnerable, both to warming temperatures and declining oxygen levels. That's especially true in the live lobster shipping trade, where it's really important to keep those temperatures cool and oxygen levels high, in order to be able to increase the probability that you'll have really high survival of those high-priced lobsters that you're shipping to Asia or the U.K."

Especially troubling for biologists watching this warming trend is a phenomenon called dead zones, low-oxygen areas of the sea that used to be teeming with marine life but are essentially becoming biological deserts. In some shallow southern New England waters, these areas are becoming graveyards for lobsters. Scientists from the Woods Hole Oceanographic Institution, a nonprofit organization in Massachusetts focused on ocean research, exploration, and education, think algae growth caused by warmer water contributes to this fatal lack of oxygen and say the problem is pushing lobster into new places and forcing fishermen to move their traps to follow them. In the Cape Cod area of Massachusetts, the state has implemented a monitoring system; fishermen can upload data on dissolved oxygen levels to help them avoid these low-oxygen patches, called hypoxic zones.

"It's another concerning feature that we're starting to see," Rick says. "Throughout the southern part of Cape Cod Bay, and just on the northern side of the arm of Cape Cod, over the past few years we've been seeing a dead zone. Monitors in the fishery in that area have been pulling up traps with dead lobsters in them."

It's clear that the age of abundance in the former heartland of North America's lobster fishery is ending, but it's still difficult to accurately guess how many lobsters there are in the ocean, or will be in the future. On the Canadian side of the border, fisheries biologists are still trying to determine the maximum amount of fishing that the population can sustain.

Canadian fisheries managers are closely watching the declines in the harvest in the U.S., but still have an incomplete picture of what's happening at the bottom of the ocean, says Rémy, the University of New Brunswick biologist. That's why it's important that scientists like him work with fishermen to help monitor lobster stocks. Many younger fishermen welcome this research, Rémy says, though he concedes that some old-school lobstermen are wary of cooperating with scientists, worried that more monitoring will ultimately mean more restrictions on their fishery.

They're not entirely wrong. But scientists argue it's in everyone's interest to know what's happening to lobster populations so the fishery can be better managed and sustained for generations to come. Without appropriate restrictions in place, biologists worry that overfishing is only speeding up the decline being caused by a warming climate. Fisheries regulators in both Canada and the U.S. have the power to restrict the number of traps and licences and enforce minimum and maximum legal sizes, but their hands are tied in one key area.

With no quota system to control the total volume of lobster being caught every year, Rémy says, there are few levers that can be pulled quickly to conserve stocks. Instead, things regulators have no control over—market prices, for instance, or a decline in demand in China because of tariffs—have greater ability to influence conservation of the species.

As a graduate student in the 1980s, Rick developed a simple way to measure populations of baby lobster, using a device that functions like an underwater vacuum cleaner, sucking up juvenile lobsters hiding in the cracks and crevices between rocks on the ocean floor. Once they're counted, the numbers give scientists a better way to estimate future adult lobster populations. It's critical forecasting data for North America's most valuable fishery, and every lobster fishing state and province with the exception of Nova Scotia participates in these studies.

This research is showing that the ocean is undergoing significant climate-induced changes, and these fundamental shifts mean lower survival rates for lobster larvae and, as a consequence, smaller commercial harvests. The North American lobster fishery has been through major changes before, but none quite like those many think are coming.

"I think it's fair to say," Rick tells me, "that everybody recognizes that there is this south-to-north shift happening, over a scale of decades, and the supply will increasingly be working in Canada's favour. And the onus will be on everybody to maximize the value that they can get out of this fishery, sustainably. So the industry is going to have to adapt."

7

Shell Disease

NARRAGANSETT, Rhode Island – When summertime returns to Rhode Island and Riley Secor goes back on the water to set her traps, she's looking for lobster that no fisherman ever wants to find.

Riley, a Ph.D. student at the University of Rhode Island's Graduate School of Oceanography, spends several weeks every year bobbing across Narragansett Bay, an estuary speckled with islands that used to be one of the most productive lobster zones in the U.S. She's hunting for lobsters with telltale lesions on their shells—ugly round marks that look almost like rust spots on the hood of an old car.

Rhode Island, with its ubiquitous clam shacks and sandy beaches, has become ground zero in the race to understand epizootic shell disease, a condition caused by bacteria and fuelled by warming ocean temperatures that has been worrying fishermen farther north. First observed in lobster pounds in the state in the 1930s, shell disease was considered a manageable problem until a severe outbreak in local waters in the late 1990s was linked to a massive die-off that devastated an otherwise productive fishery. The industry still hasn't recovered in Narragansett Bay, where waters are about 1.7 degrees Celsius warmer than they were a century ago.

It's a big reason the lobster fishery south of Cape Cod is widely believed to be doomed. During the 1980s, only about one in ten thousand lobsters in this area were seen with shell disease. By the late 1990s, the disease had exploded. In Narragansett and Buzzards Bay, a little farther up the coast in neighbouring Massachusetts, up to 70 per cent of the population was suddenly showing signs of lesions.

The disease, which today affects 30 to 40 per cent of Rhode Island's lobster population, has hollowed out the state's once prosperous fishery. It has never recovered from the collapse of the late 1990s, and most fishermen have either left the industry or pivoted to Jonah crab, which fetches a fraction of the price of lobster. Others have sold their traps and switched to farming oysters. Some are still trying to chase what little groundfish are left here—although those fisheries have also buckled beneath the weight of overfishing and climate change.

What scares scientists and fishermen is how shell disease has grown in lockstep with severe declines in the lobster population. In the Long Island Sound, which sits at the western edge of Rhode Island, the annual catch dropped from 1,678 tonnes in 1998 to 64 tons in 2011 as shell disease spread widely after several unusually hot summers, according to the Connecticut Department of Energy and Environmental Protection.

"The reality is, it's getting worse," says Riley, who grew up in the Midwest but fell in love with the ocean when studying in Boston. "What I'm struggling with is that, you know, if shell disease is temperature related, we can't stop temperature increase in a reasonable amount of time to make a difference. So we can't really stop the progression of shell disease. There's no antibiotic we can apply, there's no treatment we can give, there's not a whole lot practically we can do, other than follow the growth of shell disease. But I think it's really important to be prepared."

In its earliest stages, the disease doesn't affect lobster meat but makes the animal less visually appealing to consumers. But it's more than just a cosmetic problem. At its most advanced stages, shell disease saps the lobster of all energy, weakens its immune system, and slows its growth, making it sluggish and unable to defend itself. While some lobsters can shed the disease when they moult and begin growing a new shell, in many cases it's a death sentence. It particularly afflicts egg-bearing females and larger lobsters, which has only compounded concerns about declining lobster populations in the fishery's southern regions.

Rhode Island's waters are warming rapidly, even at the deepest sea levels, where temperature should be slower to change. In the fall of 2019, fishermen collecting data for the Shelf Research Fleet, a joint venture of the nonprofit Commercial Fisheries Research Foundation and the Woods Hole Oceanographic Institution, were startled to find the temperature of the ocean floor in state waters had jumped from about 10 degrees Celsius to 15.5 degrees, and stayed there for thirty-eight days.

It's common for water temperatures on the surface to change rapidly, when big storms or heavy winds churn up the water, but it's not supposed to swing that quickly 150 to 200 feet down. Mark Sweitzer, a seventy-four-year-old commercial fisherman who has been fishing out of Point Judith, Rhode Island, since 1971, says this suggests some fundamental and troubling changes to the way the ocean works.

Two decades ago, he'd never seen a lobster in these waters with shell disease, although fishermen who worked closer to shore were complaining more frequently about it. Today, it's a common, and disturbing, sight—even at fifteen miles offshore in the deep water canyons where he typically sets his traps. Lobsters who have it can be spotted as soon as they're pulled from the bottom.

"Once it's full-blown, the whole back of the shell, it's like touching a sponge," Mark says. "You don't even want to touch it."

Most fishermen throw these diseased animals back in the water, he says, but a few are sold as "old shells" at a discounted price to buyers onshore—a selective process that Riley suggests is removing the hardiest lobsters from the population and leaving behind those that are most vulnerable to disease. Mark thinks the problems began after a massive oil spill in January 1996, when a tank barge ran aground in a storm and dumped 828,000 gallons of home heating oil along the south shore of Rhode Island. The federal government sponsored a lobster restocking program intended to replace the millions of lobster killed in the spill. But fishermen like Mark think the juveniles introduced to Rhode Island, raised in hatcheries farther north, were more prone to shell disease.

The arrival of shell disease signalled the end of a great run of lobstering off the Rhode Island coast, a boom that lasted for more than two decades. When Mark started fishing, in the final years of the Vietnam War—he reasoned that being on a boat was better than fighting in the jungle in Asia—lobster was only a summertime fishery here, something fishermen did after they tied up their draggers for the season. But in the early 1980s, catches began rising dramatically. Lobstermen were suddenly catching so much that there was an influx of boats, traps, and bait into the water around Narragansett Bay. At its peak, there were close to 150 vessels working the area, and catches were outstripping those everywhere else in the country. Then, in the span of a few seasons, it all collapsed. Today, there might be fifteen lobstermen left in the area around Narragansett Bay.

Mark recalls hauling up traps that were full of lobster after only three hours on the bottom of the ocean.

"It was *extremely* productive," he says. "You could drop a hundred pots and get a thousand pounds of lobster if you wanted. And you could do that day after day. We were going out five, six, seven days a week. So if you came in on a boat like mine, with only five hundred pounds, people would be looking at you like 'what happened?' As if it was a bad day. I used to just envision that there was some kind of hole in the bottom of the ocean that they were spilling out of, like an anthill, or bees out of a beehive. It just seemed that there was so many of them."

In those boom times, tractor trailers lined up to collect lobster at the harbour in Point Judith, one of the few working fishing villages left on

this coastline, where the beaches are popular with surfers and sunbath-
ers alike. Today, the trucks don't bother coming because not enough
lobster is being unloaded here to warrant it. Mark sells the few lobsters
he catches to tourists at the dock, as part of a special program licensed
by the state to help lobstermen pay their bills. There aren't many young
fishermen getting into the business, because there's no future in it.

"I mean, my son's a helicopter pilot," Mark tells me. "This is the kind
of business where I wouldn't let him go out. He had no interest in ever
going fishing, and I would never want him to go anyway. The inshore
fishery is almost impossible to make a living at right now. Unless you're
already in it and have no debt and no family."

Fishermen have plenty of theories about why shell disease has soared
and catches have collapsed here: the introduction of disease-prone lobster; rising ocean temperatures; the arrival of new predators such as
black sea bass, drawn north by the warmer waters and with a voracious
appetite for juvenile lobster; or years of intensive fishing. Most likely,
they admit, it's a combination of all those things. But whatever is causing the once productive lobster grounds off Rhode Island to decline, it
just leaves Mark sad. A part of the state's heritage is being lost, he says.

Mark says the changes to the ocean floor off Rhode Island can't be
dismissed as "a few warm nights" that are spiking water temperatures
The Atlantic coast is being fundamentally altered, he says. Countless
studies have already shown that the water off the New England coast is
warming rapidly. Over the past thirty years, the Gulf of Maine has
warmed three times faster than the global average. For the past fifteen
of those years, it has warmed at seven times the global average. That's
making American lobster's southern range increasingly inhospitable to
the animals. It's changing the acidity and salinity of the water they live
in, slowing their growth and compromising reproduction. And, increasingly, the warming Atlantic Ocean appears to make lobster more susceptible to shell disease.

For career lobstermen like Mark, it all signals the end of a fishery that used to be a major part of the state's coastal economy.

"Maybe it'll get better in a hundred years, 150 years, who knows?" he says. "But I don't see it coming back. Not in my lifetime."

Researchers like Riley Secor know they can't prevent the ocean from warming, and they understand that shell disease can't be slowed or stopped. But they're hoping they can learn why the disease affects certain lobsters and not others, and how the fishing industry can be better prepared as the problem expands to other regions on its northward march.

Riley studies the mysteries of shell disease from inside a lab on the University of Rhode Island's seaside campus, where she monitors around forty lobsters with varying states of shell rot. They're kept apart by mesh liners that keep them from cannibalizing each other, and swim in fresh seawater pumped in from the bay. The researcher collects the specimens herself, from traps she hauls in the summer as part of a program that partners with a local fisheries association trying to monitor the impact of wind farms off the coast. In this southern edge of the North American lobster fishery, diseased lobsters aren't hard to find. They're in almost every trap.

One feature of shell disease is that it's not contagious. Instead, it's what scientists call a dysbiosis, or a bacterial imbalance. And the bacteria that cause these unsightly lesions are everywhere in the marine environment.

"It's a complex disease," Riley explains. "It's not like, for example, a cold that you catch. If you're around someone who has a cold, they can give you the cold. That's not how shell disease works. It's basically anything that can stress out a lobster, whether that be environmental, nutrition-wise, anything, really. If a lobster is left in some sort of compromised state, that can alter its immune system. And then that allows the disease to grow. Normally you'll find the bacteria everywhere in the marine environment. They exist on a lobster shell normally, and usually it's okay. But when a lobster is in some sort of compromised state, it allows that bacteria to kind of take over and create the lesions."

Yet how shell disease affects a lobster population is often misunderstood outside of scientific circles. Officials in Sweden used it as an excuse to try to ban imports of North American lobster in 2016, out of concerns that the condition could spread to domestic stocks of European lobster. It's also been raised as a concern in the U.K., where tourists accused of throwing shells overboard from seafood buffets on cruise ships were blamed for spreading the disease. But the science shows that's not how shell disease works. With the bacteria that cause shell disease naturally present in the water that lobster live in, warming water triggers something that reduces lobsters' natural immune systems.

"These bacteria exist everywhere in the marine environment. They're normally not pathogenic," Riley tells me. "But something is occurring, especially within southern New England, that is putting these lobsters in a state of compromised immunity. And so I'm trying to figure out what factors are contributing to that compromised immunity."

As water in the Gulf of Maine and along the Canadian coastline warms, the growing prevalence of shell disease is keeping some scientists and fishermen up at night. It's already showing up in around 10 per cent of lobster caught in southern Maine, and at lower levels farther up the coast. As the coastal waters in some of the lobster fishing areas continue to warm, it's expected those rates will climb toward the levels seen in Rhode Island.

Rick, the University of Maine biologist, points out, "There's a very strong correlation between warming temperatures and the prevalence of shell disease. And so it's more common the further south you go, and it's more common the shallower you go."

The most severe cases appear in lobsters that retain their shells the longest, namely, older, egg-bearing females that are the most critical to the species from a reproductive point of view. And although a stricken animal's disease disappears when it sheds its shell, it's only a temporary relief. The bacteria that cause the disease remain in the environment, and sure enough, in lobster with weakened immune systems, the telltale pits reappear. Troublingly, the disease also appears to cause blindness, a further problem for a lobster population beginning to show signs of stress.

"In Rhode Island, in fact, they were seeing such severe cases that they were causing these very extreme pathologies in egg-bearing females where they would induce moulting before the eggs hatched. Of course, that's not a good thing," Rick says.

Research suggests that lobsters expend a lot of energy as their immune system tries to fight off the disease, which explains why affected lobsters tend to grow more slowly and shed less often—a concern for a fishery where size matters immensely. And more time in between moulting means there's more time for those lesions to grow.

The relationship between shell disease severity and lobster growth is still being explored, but scientists like Riley have discovered that there's a tipping point where the most infected females are spending so much energy trying to fight the disease that they won't shed at all.

"Especially for the really, really shell-diseased females," she says, "I think there's a point of no return. I've seen this anecdotally, where if shell disease gets so bad, they won't even moult, they'll just die."

Riley has been doing her research long enough that she can spot a sick lobster just by its behaviour, not the appearance of its shell.

"They just mope around the tank," she says. "Basically, when you eat food, that's the energy you're taking in. And then you have to account for all your other bodily needs, your metabolism, your excretion, all that type of thing. Whatever energy you have left over, so to speak, is what you can use to grow. And we're finding that with shell disease, lobsters are growing at a lower rate."

Riley became interested in shell disease as a graduate student while working in the lab at URI's School of Oceanography. She wanted to know why some laboratory-raised lobsters that had never been exposed to the wild would develop the disease, despite having otherwise healthy immune systems and feeding patterns.

"What started me on this path is that we noticed that even in the laboratory, hatchery-raised lobsters that had never seen the wild, and by all accounts were living a pretty good life, were starting to get shell disease. That piqued my interest because I'm like, 'They should be having a good life. Nothing should be wrong. What am I doing wrong that they're getting shell disease?'"

Riley's research showed that nothing could be done to prevent the disease's spread, although cooler water temperatures appear to reduce its occurrence. But she and others wonder about genetics—that mysterious code that seems to protect some lobsters better than others. If immunity is built by nature into some lobsters' genes, perhaps it can be isolated and used to breed a more resilient variety of lobster in the future. Or so the thinking goes.

But for now, the future of lobstering in Rhode Island looks bleak. Riley says one solution may be to simply shut down fishing in areas where shell disease is most prominent, giving healthy lobsters a chance to multiply before they're caught. Otherwise, fishermen are selectively removing the most resilient lobsters from the ecosystem and leaving the most diseased ones behind.

"That's not a recipe for a healthy lobster population," Riley says. "I know there's not a lot of hope that we can get back to the way the fishery was before the big crash. I think the question now is, How do we manage what we have left? How can we get some sort of fishery, sustainably, out of that, knowing that it's not going to go back to what it was?"

8

—

God's Will

PORT HOOD, Nova Scotia – It was well before daybreak on May 12, 2018, when the *Ocean Star II* chugged out from behind the rocky breakwater at Murphys Pond, Nova Scotia. The crew was soon hauling in its first traps of the morning, tracking down their buoys in the dark in an early-morning ritual that has been repeated along this dramatic coastline for generations.

There was still a gentle roll to the sea, the leftover from a heavy northwestern wind that had blown through the big island a few days earlier, but the conditions were otherwise good and calm. Captain Hugh Watts, thirty-nine, his teenaged son Elijah, and deckhand Glen MacDonald, fifty-eight, were fishing close to shore—a little more than three hundred feet from the beach just north of Port Hood—in an area they had fished countless times before.

About halfway through their second string of lobster traps, as the boat's hauler was pulling another one to the surface, an unusually large wave hit the *Ocean Star II* from behind and washed over them, flooding the working deck. Hugh started to turn the vessel to meet the next wave head-on.

"It was a freak wave that just came out of nowhere and swamped the boat. We didn't see it," recalls Elijah, now twenty-three.

The *Ocean Star II* was halfway through the turn when another wave hit the boat broadside, rolling it over. Elijah had seen that one coming and dove under the washboard. Suddenly he was underwater, immersed in the numbing cold of the Northumberland Strait.

"The boat flipped right over and I kind of got all disoriented, I didn't know which way was up or down. When I came to, I was under the water," he remembers. "I swam up and hit my head, because the boat was over top of me."

By some miracle, the young deckhand found an air pocket and stayed there long enough to regain his senses. He finally pushed away from the vessel and swam up to the surface. He saw Glen crawling up on the bottom of the overturned boat.

Glen helped Elijah up onto the boat, and the pair spotted Hugh bobbing in the water about twenty-five feet away. But before they could do anything, a third wave hit them and washed them back into the water, pushing them toward the beach. Elijah swam for shore. When he reached the beach, he found Glen floating face down in a few feet of water. He pulled Glen onto land and performed CPR for about three minutes without any response. Still unable to get a pulse, and not seeing any sign of his father, he ran up the hill to a house to get help. Shortly afterwards, Hugh's body washed ashore.

Within minutes another fishing boat had arrived, and its captain drove the boat up on the beach so the crew could help the fishermen. His first mate began giving both men CPR. But it was too late. Within minutes, paramedics were pulling Hugh's and Glen's lifeless bodies up the beach.

In fishing communities like Port Hood, where the cemeteries are full of people who earned a living on the water, the risk of death has often been seen as just an implicit part of work. Many still consider the lives of lobstermen lost at sea, swept overboard, caught up in a string of traps and pulled into the water, or drowned when their boat capsized in rough water as a simple matter of fate. It's God's will. The sea provides, and the sea takes away.

In almost every corner of the world, despite efforts to improve safety across the industry, commercial fishing is one of the most dangerous occupations there is. Globally, there are more than one hundred thousand fishing-related deaths each year, or about three hundred a day. While fishing is a relatively small sector in Canada, accounting for less than half of one per cent of the workforce, it is responsible for a disproportionate share of on-the-job deaths. On average, there is one death in the Canadian fishing sector nearly every month—fatalities that often go unpublicized. There were forty-five fishing-related fatalities in Canada between 2018 and 2020, the worst three-year period in two decades. A deckhand in Canada is fourteen times more likely to die on the job than a police officer is, yet this hazardous job receives a fraction of the attention.

Nearly a dozen British fishermen died in 2021, most through drowning, in a year where there were 1,531 marine accidents. In Australia, the fishery is about twenty-five times more dangerous than work in mining and construction. In 2019, American fishermen were forty times more likely to die on the job than the average U.S. worker; an average of forty-three American fishermen die on the job each year. Nearly half die because their vessel capsized or sank, and almost a third were killed when they fell overboard.

Gail Atkinson, a lobster boat captain who fishes out of Lunenburg, Nova Scotia, thinks part of the reason fishing fatalities and accidents remain so stubbornly high is because of the increased pressure fishermen feel to push themselves to the limit and cash in on our endless demand for more seafood. Canadian lobster exports topped $3.2 billion in 2021, a historic record and a $700-million increase over pre-pandemic levels, according to federal trade data.

Prices, while they do fluctuate, have grown steadily over the years as record harvests have been landed. Canadian inshore fishermen caught an estimated 103,087 tonnes of lobster in 2021, according to the Lobster Council of Canada—a new industry peak that's more than twice the amount landed just fifteen years ago.

Historically high prices have also spawned a gold rush mentality to pluck as many lobsters from the bottom of the ocean as possible. It's motivating some fishermen to take bigger risks, she says. They're going out to sea longer, in rougher weather and on little sleep. That financial incentive, and the growing debt load for many new fishermen who are buying increasingly expensive licences and boats, is also hurting efforts by the federal government and fishing associations to improve safety and training in the industry.

"I feel like we're taking bigger and bigger risks," says Gail, a plain-spoken fisherman's daughter who grew up around the rhythm of the tide in Cape Sable Island, Nova Scotia. "We're pushing ourselves out there way more than our grandfathers and fathers would have ever done. We're going harder and harder, for the money."

In her homeport, a hilly tourist town famous for being the birthplace of the *Bluenose* racing schooner, a memorial near the historic waterfront lists the names of the more than six hundred local fishermen who have lost their lives at sea since 1890.

As a trailblazer in an industry still dominated by men, Gail is trying to change old stereotypes about lobster fishermen as saltwater cowboys who regularly risk their lives for their prized catch. On her boat, the *Nellie Row*, she has strict rules for moving around on deck, and is always monitoring her crew on cameras—she even wants them to let her know when they're going on bathroom breaks.

Matthew Duffy, executive director of Fish Safe NS, an association that promotes safety training in Nova Scotia's commercial fishing sector, believes financial stress is a factor for a lot of fishermen who make risky decisions. He points out that it can cost a fisherman more than $1 million Canadian to buy a lobster licence and a boat in his province, debt that can put pressure on some to go out fishing on days when it's not advisable.

"When we're looking at record-high prices, there are circumstances where some folks will push the envelope more than they would in the past," he says. "For someone who's twenty-seven, thirty years old, that's a lot of financial stress to take on. And the way you pay it off is hoping there's something in your traps."

But Matthew insists the culture is changing within the industry. Young fishermen don't question wearing life jackets in the way older generations used to, he says. Workers' compensation claims for injuries have also been dropping among fishermen in recent years, proof, he says, that training and improved protocols are working.

"We really don't face any resistance anymore when it comes to talking about safety," he says. "It's still a fine balance. But we need people to understand the risks when they decide to go out in bad weather. It's not just the loss of a catch. It could be the loss of life."

Gail knows that danger will always remain an inherent part of the job. She feels personally responsible for the safety of the all-female crew who work for her. In the bone-chilling water around Nova Scotia, she also knows life jackets and immersion suits alone can't protect a fisherman from death. Two days before she spoke, a fisherman from Digby, Nova Scotia, died of hypothermia after the scallop boat he was working on sank in the unforgiving waters off eastern Nova Scotia. The man had spent five hours in the Bay of Fundy, where water temperatures seldom exceed 8 degrees Celsius even at the height of summer, before he was found by the coast guard. He died despite wearing an immersion suit designed to keep him afloat and protect against the cold.

"A captain's worst nightmare is to lose someone at sea," Gail tells me. "The burden of someone else's life is huge. From the time I get in my car and drive to the wharf, that burden is always on my head."

When the *Ocean Star II* capsized, no one was wearing life jackets or any other type of life-saving device, although four personal flotation devices, or PFDs, were recovered on the shore, according to an investigation by Canada's Transportation Safety Board. But Elijah Watts insists life jackets wouldn't have helped anyone that day. Instead, he believes they would have trapped the fishermen under the boat, making it harder for them to free themselves from the capsized vessel.

Life jackets remain a rare sight on many commercial fishing boats. They're often there, stowed away, but many fishermen still choose not to wear them. Some argue they make it hard to manoeuvre and work on the deck; others complain PFDs won't save them from hypothermia anyway. Although those old attitudes are changing, say people who promote safety campaigns in the industry, there's a fatalist bent among those who earn a living on the sea that still endures.

Within minutes of Elijah's 911 call, fishermen all along the western coast of Cape Breton island spread the news on the local channel of their VHF radios. One of them was Iain MacEachern, who fishes out of Baxters Cove, less than twenty minutes by car down the coast. His son, a friend of Elijah's who played hockey with the teenager, was fishing with him that day.

"Everybody heard about that pretty quickly," he tells me, over the din of the aerator in the Port Hood fisheries co-op's lobster tank. "I just knew it wasn't good."

Iain now fishes full-time with his son, and admits he thinks more about the risks of commercial fishing than he used to. Returning to the water just two days after the accident—most would have gone out the next day, but it was Sunday, a traditional day off for fishermen in these parts—was grim for many in the community, he says. But it was still mid-season, and they had a job to do.

"The next day, you go out again, and you're thinking about it. But you've got to go to work," he insists. "I used to not worry about it, but now that my son's with me, it's a little bit in the back of your head. Fishing is a hard job. People think you just jump in your boat and go."

The fleet that called Port Hood's harbour home was small, perhaps a dozen boats. Everyone along this coastline would have known the *Ocean Star II*, and its crew. Hugh Watts was a volunteer firefighter, and

Glen MacDonald had been fishing in the area for years. Hugh was known for his sense of humour and his devout Catholic faith, insisting his family say the rosary with him every day. Glen, an auto mechanic by trade, was one of those people who seemed to be able to fix anything.

Soon, word of the accident crackled over radios and among neighbours. People drove to the wharf and lined the shore, hoping and praying the crew would be found alive.

"The whole town just stopped," recalls Bernie MacDonald, manager of the Ceilidh Fishermen's Co-op in Port Hood, which Hugh was a member of; he bought his bait, rope, and traps from the co-op and sold them his catch. "It affected everyone, not just the fishermen on the water. Everything slowed right down."

Hugh was buried a few days later below St. Peter's Catholic Church, on the hill overlooking Port Hood Island. His black granite gravestone reads "Let it be done to me, according to thy will." The big church stands tall on the hill, like an imposing red-brick anchor keeping Port Hood from slipping into the sea. The sign above its cemetery, in use since 1800, says "Pray for us." Hugh's classmates from high school pooled their money and installed a marble bench by the harbour in Murphys Pond, inscribed with the words "Hail Queen of Heaven the Ocean Star," taken from an old Catholic hymn full of nautical imagery.

In small fishing towns like Port Hood, many accept that fishermen are sometimes lost at sea. But that doesn't mean it doesn't leave a scar on the families most directly affected. The *Ocean Star II* capsized the day before Mother's Day, when the crew were out trying to bring home some lobster for supper.

"It's been four years, and I still can't go there," says Betty Watts, Hugh's mother. "It's just devastated our family."

Hugh had raised Elijah as his own son after he met the boy's mother in North Carolina. Elijah was on a boat almost as soon as the young family moved to Cape Breton, getting home-schooled by his parents so that he could work full-time by the time he was in grade 6.

Slowly, life got back to normal after the accident. Elijah took just four days off—passing through the funerals and visitations in a blur. Things didn't feel right, he says, until he went back on the water.

"I wasn't at peace with myself just sitting at home. I had to get back out on the water," he tells me.

Elijah today fishes in memory of Hugh and Glen, following the lobster, tuna, and crab seasons throughout the year. He says it's important to keep going to honour his father's legacy—celebrating a man who was widely respected around "the pond," as Port Hood's wharf is known. The idea that a fisherman, born to do this work, might choose to stop fishing was out of the question.

"I've been doing it my whole life, and I don't know where else I'd go," he says. "I just wanted to do right by my dad. This is the only thing that I knew to do."

9

Dying to Fish

MACES BAY, New Brunswick – Brad Small is standing in his kitchen, gulping down his fourth cup of coffee of the day, and talking about dying. He speaks admiringly of an older man down the road who had a fatal heart attack while working on a scallop dragger, took his last breath of sea air, and fell overboard.

"What a way to go," he says with a smile. "That's as good as it gets." Fishermen like Brad want to fish until their bodies give way, and then they want to be immediately buried in the cemetery on the hill. There is no other life for them. Retiring, or "coming ashore," as Brad calls it, has zero appeal.

"The way I see it, when you come ashore, you're either going to one of two places. The nursing home or Brennan's," he says, referring to a funeral home in nearby Saint John.

To him, there is nothing worse than a fisherman who can't do the work anymore but keeps on living. He talks about how his father, after he lost the ability to climb aboard his boat himself, always wanted to be hoisted up onto the deck, just to sit there awhile. In the world of the lobsterman, hard work has always been a point of pride. They are out hauling traps at sea long before most people even wake up in the morning. The front of Brad's boat serves as a reminder, with "Genesis 3:19" painted on for everyone to see.

"Man is to eat by the sweat of his brow," Brad says, quoting the verse.

But no matter how hard he works, he can't slow the effects of the changing ocean unfolding rapidly just a few yards beyond his front door. From the big bay window that faces the water, Brad can see the treeless expanse of the Salkeld Islands, two flat streaks of land named after the Quaker settler who first owned them in the nineteenth century. As a boy, he used to pick dulse, a purple seaweed used as a salty snack and health food additive, and periwinkles in front of the islands at low tide. That's how he paid for his first bicycle, he says proudly. But the dulse hardly grows on those same rocks anymore. One theory is that the bay has become too warm for the delicate seaweed to grow in places it used to be abundant.

There's little time to dwell on that right now. When he finishes his coffee, Brad is back on the phone, trying to confirm with his deckhands that they'll be heading back out on Wednesday. Today, his boat will stay moored at the wharf, as a powerful storm tears across the region, whipping the bay into a brown fury and tearing the walls off a nearby lobster pound still under construction.

"It'll blow the rabbit right out of the woods," he says. "It's a dirty, dirty day."

With the boat tied up behind the safety of the breakwater, there's still plenty of work to do. Heavy jugs of motor oil need to be fetched from a nearby marine supply shop. There are traps to be mended and fishing lines to be fixed. Mechanical issues are constant, and when they're not on the water, Brad and his first mate Tom Duke are working on the boat's big diesel engines and network of pumps that control ballast and circulate seawater in the vessel's live well. Brad, as the boat's captain, isn't above any of these tasks. When it's time to fill fresh bait bags, which is done by hand-scooping a slurry of foul-smelling fish from a giant plastic bin, he sits down and starts packing those, too. It's smelly work that seeps into every pore in your skin. When Brad gets home, he strips off his navy blue hoodie and throws it on the kitchen floor. He jokes that a smell like that can't just be washed out. You need to soak it in vinegar.

"Take that outside and burn it," says his wife, as she arrives home after a shopping trip in town.

Brad smiles and feigns protest. "Those smells, to me, are memories," he says.

When Brad was four, his father began taking him to work with the other men down on the wharf. They taught him everything a fisherman needs to know about boats and working on the water, and how to chew tobacco and smear sardines on a slice of bread. They teased him constantly, but he loved the attention. At age five he was working a pair of homemade traps in the harbour; the older men warned him he'd be arrested for fishing without a licence. In those days, before the government restricted access to the fishery, anyone could buy a lobster licence at their local general store for fifty cents. Brad still regrets not buying one. But nothing made him prouder than when his father would show up at school to take him out of class to go fishing for the day. He'd walk out of the building a little taller than the rest of the kids, he says.

"My wife says I didn't have a childhood because I spent all my time with the men. But I wouldn't trade it for anything," he tells me. "Fishing is more than just what I do. It's *who* I am. This has been my whole life. If I have to give this up, I don't know what I'd do. That wouldn't be good for me."

Understandably, Brad feels strongly about anything that threatens his way of life and his livelihood. But he's the first to concede that fishermen also need regulation. Without limits, he says, they'll catch everything in the bay until there's nothing left. The need for conservation is why there are open and closed fishing seasons, and restrictions on the number of traps, type of equipment, and lobster sizes, he explains.

"This is a competitive industry. Nobody wants to catch less than the next guy, or less than they did last year," Brad says. "I'm like a hockey player whose job it is to put the puck in the back of the net. And I'll fight and scrap and do whatever it takes to do that. That's why we need referees. We need regulation, to slow me down."

While many commercial lobster fishermen accept that declining catches are part of their future, there is growing concern about the arrival of new, and unregulated, competition in their industry that they complain isn't playing by the same rules. The rise of a commercial-scale fishery run by Canada's Indigenous people, and the fight over who has the right to catch lobster, is becoming one of the most complicated and emotional issues to ever face the industry. And it's pulled people like Brad into a battle over Canada's history of mistreatment of its First Nations and how modern governments should interpret treaties that were written long before Canada was a country.

Simmering for decades, the issue boiled over in the fall of 2020 when Mi'kmaq in Atlantic Canada began exercising their right to catch lobster out of season and without federal licences, and started selling the shellfish on the black market. Angry mobs of mostly white fishermen fought back against what they saw as an illegal fishery, shooting flares at boats on the water, barricading Mi'kmaq fishermen, burning vehicles, and ransacking a warehouse. Brad, who is head of the Fundy North Fishermen's Association, worked with other fisheries groups in the region to try to keep their members calm, but said some fishermen began carrying guns on their boats, before cooler heads finally prevailed.

Fishermen in this area are known to put up a fight when they feel someone is encroaching on their territory. Not far down the coast from where Brad fishes, fighting between American and Canadian fishermen has periodically erupted over a disputed patch of fishing grounds known as the "grey zone." By a strange accident of geography, the water around tiny, treeless Machias Seal Island is claimed by both countries. As lobster catches declined farther south and prices soared, more Americans began fishing here, leading to confrontations with Canadian fishermen who feel the lobster grounds belong to them. There have been gunshots, cut trap lines, sabotage, and death threats. One Maine lobsterman even had his thumb ripped off as he was trying to free his equipment while jostling with a Canadian for space. The Americans complain that the Canadians don't need to compete with them for a summertime fishery in the grey zone because they still have a winter fishery they can go back to, but they do it anyway.

"Canadians are like Vikings. They'll rape and pillage and not give a shit, because they can still go home," John Drouin, a longtime fisherman in Maine, told *Maclean's* magazine in 2015. "Somebody is going to get killed."

What concerns Brad now is that the Indigenous fishing rights issue could be creating another international dispute over lobster. Tribes in Maine have been watching closely and eyeing the growing Indigenous fishery in Canada. In 2021, Canada's Supreme Court made a landmark ruling that opened the door to Indigenous fishermen from the U.S. coming north to fish in Canadian water.

That was the Desautel decision, named after Richard Lee Desautel, a member of the Lakes Tribe of Washington State who was charged by Canadian authorities after he shot and killed an elk in British Columbia. B.C.'s Wildlife Act prohibits hunting without a licence and hunting by nonresidents. But the American argued he had an aboriginal right to hunt in the area where he shot the elk, because his people's traditional territory predated the international border. The Supreme Court agreed, and dismissed the charges against him.

The decision encouraged some members of Maine's Passamaquoddy Tribe to enter the Canadian glass eel fishery, a lucrative harvest with baby eels fetching as much as $3,000 a pound. Some of the American Indians began travelling to Nova Scotia rivers to set their nets, arguing they too were exercising a traditional right that their ancestors held long before the arrival of Europeans. That only added to tension over the exploding fishery, which was temporarily closed by federal officials in 2020 because so many Indigenous fishermen were flooding into waterways in eastern Canada.

Although lobster boats from the American tribe in Maine have yet to venture into Canadian waters, Brad is worried that's coming. He's frustrated that commercial fishermen have no say in legal decisions that directly affect their industry. And he's concerned there could be more violence on the water as fishermen begin to literally fight over the same patches of ocean floor.

"If that were to happen, it would get ugly," Brad says. "I don't know how we'd be able to hold guys back."

The same week that Brad and the crew of the *Small Fortune's* started their fall lobster season, fishermen from four Nova Scotia First Nations communities also took to the water and began setting traps, under a legal principle called a moderate livelihood fishery. This new fishery, with its roots in treaties signed centuries ago between Canada's colonial governments and the Mi'kmaq people, is still small compared with the commercial lobster fishery—with fewer than eleven thousand traps used by Indigenous fishermen in 2023, less than 1 per cent of what the commercial fishery uses. Several thousand more traps are being used by bands that refuse to cooperate with government regulators, fishing for lobster out of season and without federal licences, openly defying the Department of Fisheries and Oceans (DFO). Some Mi'kmaq people do not accept Canada's insistence that any fishing under treaty rights must take place during federally designated fishing season.

On Brad's side of the bay, in New Brunswick, more Indigenous fishermen are beginning to enter the fishery, too. One reserve, the Sitansisk (St. Mary's) First Nation, wanted to build a seven-thousand-square-foot lobster holding facility with two holding tanks, big enough to keep eighty thousand pounds of lobster, a proposal that was opposed by the residents of nearby Chamcook Harbour. For an increasing number of reserves around the region, the fisheries, and particularly lobster, are seen as an opportunity they can't pass up.

The scale of Canada's Indigenous lobster fleet may still be relatively tiny, but what concerns fishermen like Brad is where it will stop. The Assembly of Nova Scotia Mi'kmaw Chiefs is pushing the federal government to greatly increase the number of traps allowed under the moderate livelihood fishery. Between July 2022 and October 2023, Mi'kmaq fishermen caught 193,273 pounds of lobster, according to the chiefs, still a fraction of the catch of their counterparts in the commercial fishery. Yet Brad worries that if every one of the more than sixty-seven thousand Mi'kmaq in Atlantic Canada claim their right to a moderate livelihood, or lease that right to someone else who does the fishing for them, the impact on lobster stocks would be staggering.

"How big is it going to get?" he says. "Does every Native person just get to start lobster fishing?"

A common claim from commercial fishermen is that the arrival of First Nations harvesters in larger numbers will destroy the fishery. To support that view, the Coldwater Lobster Association points to federal data showing a notable dip in the amount of lobster brought ashore from St. Marys Bay, a Nova Scotia sub-basin of the Gulf of Maine that's been at the centre of this contentious fishery since 2017. The association contends that out-of-season fishing by First Nations caused the drop. But Mike Sack, the former chief of Sipekne'katik First Nation who launched the out-of-season fishery, says claims of Indigenous overfishing are not only inaccurate but only lead to more conflict against his people. He says commercial fishermen take a far greater share of the lobster and should look no further than themselves if they're concerned about overfishing.

At stake in this debate is Atlantic Canada's most valuable fishery and most important seafood export. Lobster landings in Atlantic Canada have been worth as much as $2 billion per year in recent harvests, according to federal figures. That's more than double the total for cod, haddock, halibut, herring, hake, eel, tuna, mackerel, swordfish, shrimp, and scallops combined.

Indigenous fishing ventures, often called treaty fisheries, are widely seen as part of the evolving relationship between Canada and its original peoples, part of a process of reconciliation for the racist legacy of colonialism. Brad seems to understand that that history needs to be addressed. But he's frustrated at the way his industry has been left out of most of the policymaking, and worries there won't be much of a lobster industry left to leave to his children if the burgeoning treaty fishery is allowed to grow unchecked. So far, he says, Canada's courts have excluded the viewpoint of fishermen like him while making decisions that directly affect their livelihoods. The federal government has also spent hundreds of millions buying equipment, fishing boats, and access for Indigenous fishermen. Brad says that for non-Indigenous fishermen who've paid $800,000 or more for a lobster licence and are drowning in debt, it's infuriating watching someone start competing with you for the same lobster without having to pay for a licence or a boat and without needing to abide by the same restrictions around when they can and can't fish.

"Legally, we don't have any say. We're shut out of all these decisions," he says. "It would be as if someone just came into your house and started taking your stuff. It's time to take a stand."

PART TWO

CONFLICT

10

There Will Be Blood

ISLE MADAME, Nova Scotia – After Phillip Boudreau's lifeless body drifted out of sight, sinking down into the salty depths, the fishermen went back to doing what they did most mornings during lobster season—they continued hauling their traps.

A deckhand and skipper of the *Twin Maggies* had just shot, gaffed and dragged a man from their village out to sea for a crime that is often settled without the police in rural fishing communities—stealing their lobster and damaging their gear. On the morning of June 1, 2013, under a sparkling clear blue sky, the Nova Scotia lobstermen were working in a small cove known as l'Anse aux Maquereaux, or Mackerel Cove, when they spotted Phillip's smaller speedboat, the *Midnight Slider*, moving among their traps. After years of having their lobster gear vandalized, and often hearing Phillip boast about it, the fishermen had seen enough. According to the account given during their trial, deckhand James Landry fired at Phillip with a shotgun, as captain Dwayne Samson rammed his smaller boat, partially submerging it, while James hooked him with a long gaff and held his body under water until an anchor could be secured around his neck. Then they dropped him to the ocean floor.

"You won't cut any more of our traps," said James, one of three men onboard, according to court records, as Phillip's body sunk into the sea.

On Isle Madame, a French-speaking enclave on the rocky eastern edge of Cape Breton, Phillip was widely known as a thief and a bully and was hated among many fishermen for his behaviour. The cut traps had been costing the owner of the *Twin Maggies* between $5,000 and $6,000 a year. But his killing stunned even those who despised him here, and made headlines around the world. The case was dubbed "Murder for Lobster," and it shed a light on the problem of frontier-style justice that sometimes pervades the fishery. Violence, while still rare, flashes up every season around North America's lobstering regions. As the oceans warm, as lobster populations shift away from historical grounds, and as fishermen begin to worry more about their livelihood—and who's taking it away from them—more conflicts are to be expected.

The killing of Phillip Boudreau was a particularly chilling example. In and around Arichat, the largest village on the island, Acadian pride is on full display, with the red, white, and blue colours of the Acadian flag painted on telephone poles, boats, and mailboxes. It's a constant reminder of the distinct heritage of the people who call this place home. Hundreds of small, tidy homes are tucked into the rocky shoreline, wedged between the treeless scrubland and the Atlantic Ocean. In every community around the island, the white-stained Catholic churches are the tallest buildings in town. The oldest Catholic church in Nova Scotia, the cathedral of Notre Dame de l'Assomption, built in 1835, stands like a fortress with its twin bell towers on a hill overlooking Arichat's harbour. The fishery may be the only thing that competes with the church for reverence here. That, perhaps, and the beloved Red Caps, the local baseball team who play in nearby Petit-de-Grat's charming ballfield, which sits so close to the sea that a long home run over the right field fence just might splash into the ocean.

On this island, taking away a person's ability to fish is one of the deepest, and almost most sacrilegious, of offences. It's been that way for centuries. It was the rule when the villages here were sacked by John Paul Jones, an American privateer, in raids during the American Revolutionary War. The goal of the attackers was to destroy the fishery—not so much lobster in those days as codfish, which was dried and salted on Isle Madame. At the time, Arichat was one of the most important ports in Nova Scotia, and it was a bustling little hub of shipbuilding and fishing.

The Acadians who live here have long memories. Nova Scotia's French-speaking minority had to learn how to survive by fishing in the most unwelcoming of places, after the British kicked them off their land in the Annapolis Valley and elsewhere during the war between France and England for control of Atlantic Canada in the 1700s. Stuck in the middle of the conflict between those two colonial powers, thousands of Acadians were deported and dispersed around the globe in what's called the Great Expulsion, and resettled in such places as Louisiana, Massachusetts, France, and the Caribbean. Thousands never survived the trips. The hardiest among them migrated to pockets of isolated coastline in the Maritime provinces that were unwanted by the British, and there they became skilled fishermen. In some Acadian communities, that history has fostered a culture of self-reliance, and an understandable suspicion of Canada's British-style legal system.

By 2011, Phillip Boudreau had a well-earned reputation in his community as a thief and a poacher. He regularly stole his neighbours' all-terrain vehicles, boat motors, and fishing gear. Anything that wasn't bolted down had a habit of walking away in the night when Phillip was around. Yet he often seemed to evade police—sometimes jumping off the wharf and swimming across the harbour to get away, according to author Silver Donald Cameron's book about his murder. Phillip always seemed to be on the run from something. He spent many nights sleeping in the woods, under boats, or on the open moor or hiding in trailers to avoid being caught. Once, he swam out to a small island in the harbour and hid among the rocks there, jeering and laughing at the RCMP officers when they couldn't find him, Silver wrote.

The morning he was killed, Phillip put up a fight to save his life. After he was shot in the leg, his boat was rammed and he was tossed overboard. As he flailed around in the water, he clung to a red gas can to stay afloat. James, the deckhand, hooked him with the gaff and told Dwayne to push the diesel engine and start dragging him out to sea. But Phillip wriggled out of his sweater, and the *Twin Maggies* turned around again. This time, the gaff took firmly, and James held Phillip underwater as the boat steamed away until foam began to bubble up from his mouth, the court heard. James and Dwayne wrapped a line under his arms and around his neck, attached it to a heavy stainless steel grapnel, and dropped the body into the water at a depth of about twelve fathoms. Then the crew of the *Twin Maggies* towed Phillip's damaged speedboat away and tried to swamp it. After they had unloaded the day's catch at the Premium Seafoods wharf in Arichat, James snuck the rifle used to shoot Phillip off the boat, concealed in a blanket, and Dwayne tried to scrub away marks on his hull where the two boats had collided. Afterwards, James claimed Phillip had come roaring out of the fog and run into them and they left him stranded in the water. A jury didn't buy the story. James was handed a fourteen-year sentence in federal prison for manslaughter; Dwayne was sentenced to ten years for his role in the killing. The other deckhand on board, Craig Landry, was convicted of accessory after the fact and was given two years parole.

Some on Isle Madame saw justice in Phillip's killing. Others, who knew another side to Phillip, the one who would give lobsters to older neighbours or bring deer meat to those who needed it, argued that Phillip was a tragic villain, the victim of a violent childhood and the kind of wretched rural poverty that used to plague Cape Breton. Even as an adult, he had no bedroom, and slept on a mattress in his mother's kitchen. No good came from his death, they say, and it divides his small community still.

As shocking as the killing in Cape Breton may be, brutality on the high seas is as old as fishing itself. It's a particular problem in distant-water fishing fleets, where crews operate a long way from the oversight of regulators. In one particularly troubling case in 2013, members of a Taiwanese-flagged tuna-fishing vessel shot and killed at least four fishermen floating helplessly in the Indian Ocean—a grisly murder discovered only because a shaky cellphone video taken by one of the attackers was left in a Fijian taxi a year later. Violence that happens far out at sea often happens with impunity. At least four ships were at the scene of the shooting, but there were no laws requiring any of the dozens of witnesses to report the killings.

China has the world's largest long-distance fishing fleet and is one of the worst offenders when it comes to crime at sea. Its privately owned vessels regularly fish in African, Asian, and South American waters, far from sight, typically relying on migrant crews who report frequent human rights violations. A report by the Environmental Justice Foundation, a London-based organization that is working to end illegal fishing, says roughly 58 per cent of Indonesian crew members who worked on Chinese vessels had seen or experienced physical violence while on the water.

In the North American lobster fishery, violence is not the norm for most fishermen, even the most independent-minded. But a vigilante streak runs through the industry, and it has been there since the very beginning. In these rural communities, the police are often far away and under-resourced, so fishermen learn to handle problems on their own. In Grand Manan, New Brunswick, people still talk about the night in 2006 when an angry mob of lobstermen ran a crack cocaine dealer off their island, burning his house down; the RCMP had to escort him to safety. When the police do show up, the culture of silence among fishermen makes it hard to solve these crimes.

Most lobstermen understand and respect long-held claims to individual fishing territory, but disagreements do sometimes flare up. And when they do, lobstermen have long protected their turf with threats, intimidation, and violence if necessary. Usually, these unwritten rules are enforced through sabotage, such as cut trap lines, stolen buoys, and boats that mysteriously sink in the middle of the night. But in some cases lives have been put at risk. The violence that erupted in Nova Scotia in 2020 in opposition to the Mi'kmaq fishermen's commercial fishery drew international attention, but those kinds of conflicts continue on a smaller scale, often without notice beyond the rural communities where they occur.

This territoriality is a feature of the fishery in both Canada and the U.S. In Maine, anyone with a lobster licence can set traps anywhere in one of the seven zones that divide state waters. With most lobstermen on an eight-hundred-trap limit, that can make for a crowded ocean floor—nearly three million traps are in use in the state's waters. Unofficially, each harbour sets its own boundaries, often determined by local lobstermen themselves based on decades of fishing habits. Outsiders, either from out of state or without a family connection to the nearest community, often find their arrival is less than welcomed.

In the rich lobster grounds around Matinicus Isle, about twenty miles off the coast of Maine, there's an unofficial rule that only islanders can lay down traps—even though mainland fishermen who are licensed can legally fish in the same water. Families who have staked their claim to specific sections of the ocean's floor here say this understanding has been around since the nineteenth century. Some of the island's fifty-three residents, about half of them full-time lobstermen, are willing to defend their turf by any means necessary.

"The informal rule is, if you want to fish here you have to live on the island," Jim Acheson, an anthropologist who has written extensively on Maine's lobster industry, told *Soundings* magazine in 2009. "It has been that way at least as far back as the 1890s. It may go farther back than that."

Jim's work has often focused on the phenomenon of "harbour gangs," informal networks of fishermen from the same port who claim and defend their fishing areas from other fishermen. This kind of territoriality is unique to the lobster fishery, he says. No other fishery polices who can and can't fish in certain areas quite like this. Understandably, established fishermen don't want to share their lobster grounds with more fishermen, especially as catches trend downward. Novice fishermen who only work part-time and who encroach on the space of more experienced lobstermen, sometimes entangling their trap lines, are especially unwelcome.

On island fishing communities, where entrance to harbour gangs is tied to local land ownership, outsiders find it almost impossible to break in. In 2009, a festering dispute between two fishermen near Matinicus led to one being shot in the neck and nearly bleeding to death before he could be taken to hospital. That season, lobster prices plummeted to the lowest point in decades, and fishermen were tense and anxious about higher costs for bait, fuel, and gear.

Into that environment steamed Alan Miller, a mainlander who began fishing within the invisible boundary that another fisherman considered his own. In July of that season, Alan's father-in-law Vance Bunker, a veteran lobsterman on the island, was accused of shooting Chris Young in the neck with a .22 caliber pistol, after the two had reportedly argued for weeks over whether Alan had the right to fish near Matinicus. In the trial that followed, the court heard that Vance and Chris had quarrelled earlier in the day. Vance was acquitted by the jury; he told the court he had pulled the trigger in order to protect his daughter. That same month, three boats in nearby Owls Head sunk overnight under suspicious circumstances.

Major John Fetterman of the Maine Marine Patrol, who was in charge of the state's fifty enforcement officers during the dispute, says there's simply too much coastline, and too many fishermen, for the government's agents to effectively enforce the law. So a form of frontier justice has evolved to keep rivals in check.

"They become the judge and jury with regards to who can fish and who can't," he told journalist Jim Flannery after the shooting on Matinicus. "All these invisible lines are drawn within the community about who can fish, when they can fish, and the state doesn't recognize them. Yet the territorial lines are so deeply rooted in these communities."

In 2016, another trap war erupted in Maine between fishermen who work neighbouring lobster zones near Hancock County, home to Bar Harbor and Acadia National Park. More than $350,000 worth of gear was destroyed by rival fishermen in a bitter feud that spurred the Maine Marine Patrol to offer a $15,000 reward for information that led to arrests. But silence, once again, stifled the investigation.

On Matinicus, and the many other islands on Maine's coast where lobstering is an economic lifeblood, fishermen say they have no choice but to protect their turf. It's either that or perish.

"The idea is to make sure that people who are taking lobsters off this piece of bottom are living here on the island," Clayton Philbrook, a lobsterman whose ancestors settled here in the 1820s, told *The New York Times*. "If we lose control, we fold up and die, that's it."

Before the 1990s, Maine's lobster fishery was open to anyone. Since the zone system was introduced, fishermen have tried multiple ways to keep each other out. At least one lobster zone, off Boothbay Harbor, lowered its trap limit to six hundred, which was two hundred fewer than every other zone in the state. On the surface, that may seem like a progressive conservation measure, but in reality, it serves as a disincentive to outside fishermen who don't want to fish with fewer traps.

Some make the case that these turf wars actually help conservation efforts by keeping interlopers out of lobster areas and protecting them from overfishing. Trevor Corson, author of *The Secret Life of Lobsters*, wrote in *The Atlantic*, "If all fishermen everywhere could be given reasons to feel this passionately about their long-term stake in a particular piece of ocean—within a legal framework that prevents violent confrontation—the seas might just have a chance of returning to abundance."

After the shooting on Matinicus, the island's lobstermen asked the state to create a legislatively protected fishing zone exclusively for residents. Although the state stopped short of that end, saying it needed to balance the constitutional rights of all the state's roughly 4,800 licensed lobstermen, the request wasn't without precedent. Maine approved a two-mile "conservation zone" around Monhegan Island in 1998, restricting access to local lobstermen, who had complained about interlopers from the mainland. In return, Monhegan lobstermen agreed to set fewer traps, limited to three hundred each, far fewer than the eight hundred allowed in most of the state's waters.

The New York Times reported that some people worry disputes like these, and the creation of more exclusive fishing zones, will lead to the balkanization of lobster territories along the coast. Add in the northward migration of the fishery as catches decline in more southern regions and fishermen grapple with less abundant catches from one year to the next, and the situation is ripe for contention. In the span of just a few decades, the centre of the Gulf of Maine lobster population has shifted from Casco Bay off Portland to northeastern Maine, a drive of several hours up the coast. Meanwhile, many younger fishermen are going severely into debt just to enter the fishery, and can't afford to let someone else take lobster away from them. History has shown that, when pushed to the brink, fishermen will fight to protect what they feel is theirs.

11

—

The Right to Fish

SAULNIERVILLE, Nova Scotia – The little boat is chugging away from the wharf under a dull sky when captain Levi Paul Sr., his cigarette down to a nub, flips on the speaker. Bob Marley's "Get Up, Stand Up" starts echoing through the cabin, and his four-person crew begin nodding their heads.

The tension that everyone feels back in Saulnierville, a cluster of modest homes around a treeless shoreline about thirty miles north of Yarmouth, is quickly fading as the boat closes in on its first buoy. Thirty-nine-year-old first mate Andrew Robinson hooks it with a long gaff pole, attaches the rope to the mechanical hoist, and flips the switch. Soon a trap full of squirming lobsters is pulled dripping from the sea.

Onshore, the Mi'kmaq captain and crew of the *Sea Bug* are treated as criminals by some, shunned by marine mechanics, bait suppliers, and trap makers who refuse to do business with them. But out here, on the water, they are finally free. When I showed up at the Saulnierville wharf with a notepad one breezy late-summer day in 2020 as a reporter for *The Globe and Mail*, they happily invited me to go fishing with them. They said they wanted people to understand why they were risking violence on land and at sea in order to catch lobster.

To get to the wharf every day that summer, they had to pass an RCMP checkpoint intended to keep away the commercial fishermen who say Mi'kmaq people have no right to be here. That September, their band, the Sipekne'katik First Nation, began what they called a moderate livelihood fishery—a small-scale commercial fishery that is rooted in a legal decision from Canada's Supreme Court more than twenty years earlier. For many Indigenous people, these treaty rights that give them unique access to natural resources are the foundation of their relationship with Canada as a sovereign, self-governing people.

The Sipekne'katik band launched their fishery on the anniversary of the Marshall decision, a landmark court ruling that affirmed Indigenous Peoples' right to fish, hunt, and gather in pursuit of a moderate livelihood. The entire community gathered at the government wharf to celebrate the launch, waving flags and singing songs in memory of Mi'kmaq activists who had fought to change the laws around Indigenous fishing rights in Canada decades earlier.

The celebrations soon gave way to violent protest from commercial lobster fishermen, who are mostly white and of European heritage. More than a hundred fishing vessels blocked the Sipekne'katik boats when they tried to leave the harbour, and they were chased on the water and shot at with flare guns. Police made several arrests after supporters for both sides got physical, and were called to investigate complaints of sabotage against Sipekne'katik equipment.

Commercial fishermen in Atlantic Canada, who pay hundreds of thousands for a lobster licence and must fish within a restricted season designed to control supply and prices, argue that the unregulated Sipekne'katik fishery, as small as it is, is a threat to their livelihood. The Unified Fishery Conservation Alliance, the largest fisheries advocacy group in the Maritimes was one of the loudest voices in opposition. The union says it wants "one fishery and one set of management plans" for the entire industry, not a separate set of regulations for the Mi'kmaq who fish the same waters as commercial fishermen.

Some non-Indigenous fishermen have never welcomed the sight of Mi'kmaq people on the water, arguing that the Indigenous fishery fuels a black market that undercuts their prices. They've thrown sandbags on Levi's boat in an attempt to sink it. Snowplows have pushed Mi'kmaq traps off the wharf and into the water. When Andrew's boat once broke down outside Shelburne, Nova Scotia, he says that instead of offering help, white fishermen circled him and taunted him with racist slurs.

Yet out on the protected waters of St. Marys Bay in southwestern Nova Scotia, the crew of the *Sea Bug* are hopeful for the future, proud to be Mi'kmaq and earning a living on the ocean.

"Being out here, we feel at home," says Levi, who bought his used boat with money his late grandfather left him after a long battle with dementia. "This means everything to me. We're not trying to get rich. We're doing this because it's a way for us to provide for our families."

Levi and Andrew, along with lobster handlers Nikita Paul and Kaitlin Marr and deckhand Evan Dennis, had only begun working together a few weeks earlier, but quickly settled into a rhythm. Everyone on board knows their job and goes about it efficiently. They move quickly as each trap is hauled onboard, throwing undersized lobsters back and marking females before returning them to the sea, practices that help conservation.

The *Sea Bug*, flying the red-and-white Mi'kmaq flag, is a sturdy thirty-five-foot vessel with a fibreglass hull and a hole in the deck left from an old engine fire. It's undersized by modern standards, but comfortably carries the twenty or so traps they pull up each day—a far cry from the 350 or so traps most commercial fishermen can haul up as part of their licence.

Levi, a giant of a man with a roaring laugh, has big hopes for the future. He tells me he wants to buy a bigger, better boat and one day move out of his mother's crowded house and into a home of his own. He wants his children to join him as fishermen—when it's safe again. But as happy as he is on the water, he's also worried.

Catching lobster under treaty rights is one thing; selling it is another. The Sipekne'katik fishermen struggled to find a legal market for their lobster, and the few buyers who would purchase from them were targeted by federal officials. Many of Nova Scotia's seafood buyers simply refused to purchase the band's catch, largely because of pressure from commercial fishermen. The province's wholesale buyers only recognize federally issued lobster licences, not the self-declared licences that the Sipekne'katik First Nation issued to their own people as part of their new fishery.

The Sipekne'katik fishery was seeking an exemption from the province and a change to regulation that would allow it to sell its own lobster directly to the market. For months, it kept harvesting lobsters and storing them at a facility under guard by a group of young Mi'kmaq men from Cape Breton. When I drove up to the warehouse after spending the day on the water with Levi, I was confronted by a group of them, who yelled at me to go away. I tried to explain that I was a journalist, not a fisherman, but they didn't care, understandably. For weeks they had been on guard against angry mobs who had surrounded their building, so the sight of another white man, regardless of his intentions, was not welcome. The police were stationed up on the road, watching everything. I headed back to my car and drove off.

While the Canadian federal government has tried to negotiate a settlement of the lobster dispute, it continues to take enforcement steps against Indigenous fishermen who are on the water out of season. In the two years since Sipekne'katik First Nation tried to assert their treaty right to fish whenever they want, federal conservation officers say they have seized more than seven thousand lobster traps. By the spring of 2025, the government tried a new approach, offering the Mi'kmaq band exclusive fishing zones, wharfs, and funding for training, gear, and vessels, in an exchange for a promise to only fish during the regulated season.

For former Sipekne'katik Chief Mike Sack, who faced death threats when he announced his band's new fishery, the new lobster money should allow his people to take control of their lives, buy new clothes for their kids, and break the cycle of poverty that has plagued the Mi'kmaq for generations.

"These fishermen, they become entrepreneurs," said Mike, when he sat down with me in 2020 in a hotel banquet room as the standoff continued. He was dressed like he had just stepped off a lobster boat, in a black hoodie and a ball cap that said *NDN*. "They bank it, they build their own homes, they create jobs, and that helps us collectively. It helps community morale."

Even though it's been decades since the Supreme Court ruled that Canada must recognize Indigenous fishing rights, Mi'kmaq fishermen are still fighting much of the same prejudice over their right to earn a moderate living on the water. Vandalism continues, and it has only increased their costs. Levi tells me he lost $7,000 in traps when commercial fishermen cut Sipekne'katik gear out in the bay. But for two men who have been fishing together since they were teenagers, nothing surprises them anymore.

"This ain't nothing new," says first mate Andrew. "We've been dealing with this for years. They say 'Fish alongside us.' Well, we tried that."

Twenty-six years before the *Sea Bug* was hauling lobster out of St. Marys Bay, another Mi'kmaq fisherman found himself in trouble with federal fisheries officers. In 1993, Donald Marshall, a member of Cape Breton's Membertou First Nation, was caught catching eels with illegal nets, out of season, and without a licence in Pomquet Harbour, Nova Scotia. When officials from the Department of Fisheries and Oceans (DFO) told Donald to stop fishing, he didn't comply—and was charged and convicted when he sold his 460 pounds of eels for $787. He spent six years fighting that conviction, taking it all the way to Canada's top court, which ruled he had a right to fish and hunt for a moderate livelihood as part of historical treaty rights that the British Crown had granted the Mi'kmaq in 1760.

Eel fishing for Donald wasn't just for an income. For Donald, the son of a Mi'kmaq chief, being on the water was also therapeutic and helped him deal with his post-traumatic stress disorder, a result of the eleven years he spent in prison for a crime he didn't commit. His wrongful conviction for murder at the age of seventeen in the 1971 stabbing death of his friend, a Black teenager named Sandy Seale, left him forever damaged. A rebellious youth who got into trouble with the law for petty thievery, Donald was the victim of a deeply flawed investigation and a justice system that too often saw young Mi'kmaq men as little more than criminals. He was exonerated in 1990, and his case exposed the systemic racism in Nova Scotia's antiquated courts, making him something of a folk hero among Mi'kmaq.

"The whole thing was orchestrated by a bigoted cop. It brought a focus onto just how racist Canada was," says Daniel Paul, a Mi'kmaq historian and author of *We Were Not the Savages*. "The racism was much, much worse at that time. If you were Mi'kmaq, you could be chucked in jail for pretty much nothing."

Donald Marshall's wrongful conviction helped modernize the province's courts. It prompted a Royal Commission in 1990, which made 82 recommendations to redesign the justice system, including ending the practice of appointing judges based on political connections instead of their qualifications.

The eldest of thirteen children, Donald quit school after grade 6 and soon learned that jobs for young Mi'kmaq men were hard to come by in 1960s and '70s Nova Scotia. After his conviction, his family lost a lot of work, too.

"Our opportunities in this province at that time were practically nil," says Daniel, who worked with the Union of Nova Scotia Mi'kmaq in the 1980s to help free Donald. "We've made some progress, but there's still a lot of work to be done."

The real killer, Roy Ebsary, was convicted of manslaughter in the death of Sandy Seale and sentenced to just three years in prison, a sentence that was reduced to just one year by the Nova Scotia Court of Appeal in 1986. Donald's time in prison left him with scars that never healed, his widow, Colleen D'Orsay, told me in 2020. He witnessed riots, beatings, and unspeakable violence. Many years after his release, he was still haunted by what he had seen inside.

"He used to have night terrors, he'd wake up screaming. He'd dream he was being buried alive in prison," recalled Colleen. "That conviction had a devastating effect on him and his family. Imagine being seventeen years old and realizing that your parents can't save you. He realized that the white world was not safe for Native people."

The wrongful conviction also gave Donald, who became a reluctant activist for Indigenous rights, an understandable suspicion of authority. His push for Indigenous fishing rights made him a target. Colleen recalled a time at a bar in Sydney when some men tried to start a fight with Donald over the decline of the cod fishery—an issue his Supreme Court battle had nothing to do with.

"Fishermen followed him into the bathroom and were trying to have an altercation with him," she said. "He was responsible, in these fishermen's eyes, for the collapse of the entire fishery. It was crazy."

Donald, whom those close to him knew as "Junior," died in 2009 at the age of fifty-five. When the Sipekne'katik First Nation launched their fishery in 2020, his widow was there with her son to support it. She was stunned by the violence that followed.

Donald was shy and uncomfortable in the spotlight, but fought for Mi'kmaq people in large part because of injustice he personally experienced, Colleen said. He'd be frustrated to see many of the hurdles that faced Indigenous fishermen in his era are still firmly in place, she said.

"It's déjà vu, like watching this all over again. He didn't want his children, and their children, to grow up in abject poverty. . . . He did this for his people."

The band symbolically gave its first lobster licence to Donald's eldest son, Randy Sack. The protests, racism, and violence that have followed reminded Colleen of what happened after her husband's Supreme Court treaty rights challenge all those years ago.

"Back then, he was very frustrated that the Canadian government was not holding up their end of the bargain. And sure enough, that's still happening now."

Jason Marr was nearly done unloading the last of the lobster one evening in October 2020, shortly after his band launched their controversial fishery, when he noticed a pickup truck pull a U-turn and loop back slowly in front of the fish storage plant in Middle West Pubnico, a small fishing community stuck on a peninsula that jabs into the ocean. Soon, the headlights of dozens of trucks were cutting through the foggy Nova Scotia night.

Jason, one of several Sipekne'katik fishermen at the lobster pound that night, hurried inside and locked the entrance. The men who had followed them began pounding on the door as more vehicles, carrying more white fishermen, arrived. Then rocks started flying through the windows. The crowd outside was turning into a mob, threatening those inside while someone vandalized and burned Jason's van. As the mob grew larger and larger, with more trucks arriving every minute, a terrified Jason began livestreaming.

"They destroyed my van," he says in the clip. "There's a couple of hundred out there, and I refuse to leave, I'm not leaving, they said they won't let me leave unless they have my lobsters."

The Mi'kmaq men inside were eventually allowed to leave, and the mob began ransacking the place. They confiscated around 3,300 pounds of lobster and dumped them outside while RCMP officers stood by and watched. Police said the crowd was too large to control and couldn't be stopped. Three days later, the fish plant was burned down in a suspicious fire, the warehouses on the edge of the sea reduced to a pile of blackened timbers in a matter of hours. For weeks after, Jason said the hate messages kept coming on his phone.

The RCMP eventually charged twenty-three people with ransacking the lobster pound, and two others were charged with arson. But the force was widely criticized for their hands-off approach when the violence broke out, which Indigenous leaders said only led to more lawlessness. The Sipekne'katik band, meanwhile, took its fight to the courts. It hired lawyers who collected affidavits to press police to charge people accused of damaging and destroying property, assault, harassment, and making threats.

Most of the abuse happened out of the public eye, said Mike Sack, the chief who also launched a lawsuit against the federal government for failing to protect his fishermen. Just as troubling, he said, was the effort to shut the Mi'kmaq band out of the lobster market, with commercial fishermen threatening any supplier who tried to do business with them.

The crew of the *Sea Bug*, meanwhile, told me they plan to keep fishing. For them, and many Mi'kmaq, the fight over the lobster fishery isn't just about asserting treaty rights. It's also seen as a way out of poverty.

Nearly 68 per cent of Sipekne'katik families live below the poverty line, according to a 2015 study by the Canadian Centre for Policy Alternatives. More than half of all children in Mi'kmaq communities around the province live in poverty—nearly three times the average rate in Nova Scotia. In Indian Brook, home to the Sipekne'katik First Nation, it's not uncommon for twelve people to share one 1,200-square-foot house. The community of around 1,300 residents says it has a shortage of around four hundred homes.

"We have whole families living in one bedroom," says Mike. "You drive down the road with winter coming on, and you see kids walking without a winter coat. It's that kind of stuff."

There was a time when Levi, the *Sea Bug*'s captain, didn't have many job prospects. He was a junkie, stealing five-gallon water bottles for the deposit. But since getting clean, fishing has given him a chance at a better life. As a young Mi'kmaq man with a criminal record, it's one of the few areas where he can work and employ others. He tells me that job opportunities remain limited for Indigenous people in Nova Scotia.

"All they see is the colour of my skin," he says, referring to stores around Nova Scotia that have turned him down for jobs. "Without this, my kids wouldn't have a future. I'm doing this for them."

12

—

Legal Fights

SHUBENACADIE, Nova Scotia – After four years in the U.S. Army, Michael McDonald was honourably discharged and trying to figure out what to do with the rest of his life. As a young Mi'kmaq man originally from Nova Scotia, the former soldier's attention was dominated in the summer of 1990 by only one thing: an armed standoff between Mohawk protestors and the Canadian military that would become a violent flashpoint in Canada's broken relationship with its Indigenous people.

The Oka Crisis—also known as the Kanesatake Resistance—was a nearly three-month armed standoff between Mohawk warriors and the Canadian army that began in July 1990 in Kanesatake, near the town of Oka, in rural Quebec. The protesters opposed plans for the expansion of a golf course and the development of townhouses on disputed land that included a Mohawk burial ground. The Mohawks erected barricades, and in a brief gunfight, a provincial police officer was killed. The protest didn't end until the Canadian army was called in, along with four thousand soldiers, armoured vehicles, helicopters, artillery, and police boats. In one assault, as soldiers advanced on the community, firing tear gas and warning shots while dodging rocks, seventy-five Mohawk protestors were injured along with twenty-two military personnel.

For many young Indigenous people watching across the country, the violent standoff helped politicize them as part of a growing Indigenous rights movement. For Michael McDonald, watching the protest unfold on his TV convinced him to pursue a career as a lawyer specializing in helping First Nations people in conflict with federal law.

"I saw what was happening with those First Nations people, and I thought something needs to prevent this from happening," says Michael, a large man with a broad smile and a closely cropped beard.

But Michael's story, like those of many Indigenous young men, was also interrupted by violence. His university studies were derailed in 1994 when his older brother shot and killed his younger brother during a fight in Millbrook First Nation, a Mi'kmaq reserve outside Truro, Nova Scotia. The killing sent him into a downward spiral that lasted nearly ten years, before he finally cleaned up his life and decided to go back to school. This time around, he threw himself into his school work, finished his undergrad, and graduated near the top of his class.

Since then, he's built a legal practice in Sipekne'katik First Nation, where he's the leading lawyer for Mi'kmaq fishermen who have been charged for fishing lobster out of season. And he's been winning a lot lately, successfully convincing judges that his clients, as Mi'kmaq fishermen, have a constitutional right to catch lobster beyond the reach of federal regulations. His approach, using the courts to defend his people's unique status in the law, has not won him many friends among commercial fishermen. As the First Nations fishery has grown in Atlantic Canada, that tension has only increased.

"They've had it to themselves for so long, and all the sudden, they think Native people are trying to come in and mess with their little pot of gold," Michael says. "They're screaming, 'Oh those Indians are gonna take *all* the lobster.' But we're just trying to take a tiny slice and make a living."

At first glance, Sipekne'katik seems an unlikely place from which anyone would want to launch a pioneering lobster fishery. This landlocked reserve in central Nova Scotia is more than a three-hour drive from the wharf where its members have based their fleet. A collection of modest vinyl-sided homes, it looks like many First Nations reserves in eastern Canada—a few stray dogs dodge pickup trucks that pass cannabis shops and sheds converted into Indigenous craft businesses. The red-roofed church at the top of the hill is still the tallest building in town.

First called Indian Brook by the white settlers who gathered up many of Nova Scotia's Mi'kmaq people and tried to centralize them here in the 1700s, the community changed its name in 2013 to Sipekne'katik, which means "where the wild potatoes grow." Today, while it's surrounded by rolling hills filled with corn and soybeans and dairy barns, little farming is done on the reserve lands. As the road winds toward the reserve, known as Shubie to locals, the homes get smaller, less tidy, and obviously poorer.

The telephone poles leading into the village are lined with the red-and-white Mi'kmaq flag and orange banners that say *Every Child Matters*—a reference to the national campaign to acknowledge the more than 150,000 Indigenous children who were taken from their homes and sent to residential schools in Canada between 1831 and 1996. The residents of Sipekne'katik don't need to look far for reminders of that dark chapter in Canadian history. The site of the Shubenacadie Indian Residential School is just down the road; it housed more than a thousand Mi'kmaq children between 1930 and 1966.

In this setting, far from any ocean, the community's fishery department has become the most important building in town—not the church, the school, or the band council office. The warehouse is rimmed by a tall fence topped with razor wire, as if the fishing gear and lobster traps held inside must be protected at all costs. And that says a lot about how the people of Sipekne'katik see lobster as their best chance to lift themselves out of poverty.

"A lot of these people, they had *nothing*. All they had was fishing," Michael says. "They've invested everything they have to go out and try to make a living fishing, and it's just been nonstop harassment from the Department of Fisheries."

The Mi'kmaq people of Atlantic Canada, today numbering around sixty-six thousand across the four eastern provinces, have always argued they have a legal right to hunt and fish as they please thanks to agreements signed before Canada was even a country. Sipekne'katik's claim is based on the 1752 Peace and Friendship Treaty, signed by the British governor and Mi'kmaq Chief Jean-Baptiste Cope, that recognized the right of Cope's people to catch food on their traditional territories. "It is agreed that the said Tribe of Indians shall not be hindered from, but have free liberty of Hunting & Fishing as usual," the treaty says.

Cope was chief of the tribe now known as Shubenacadie, or Sipekne'katik in the Mi'kmaq tongue, who lived on the eastern coast of Nova Scotia. The Mi'kmaq, Maliseet, and Passamaquoddy people scattered along huge stretches of the Maritimes and the Gaspé Peninsula signed around a dozen of these peace and friendship treaties in the eighteenth century. Indigenous scholars have long argued that these deals, signed over the span of several decades, should be taken together as a "great chain of agreement" that established their right to harvest natural resources.

The irony of today's dispute over Indigenous fishing rights in eastern Canada is that the treaties they're based upon were born largely out of concern that European and New England fishermen were increasingly fishing in First Nations territory in the early 1700s. As more settlers moved northward and began fishing in coastal areas and rivers that the Mi'kmaq had fished for several thousand years, violent confrontations were increasingly common.

Britain, which was vastly outnumbered by the tribes and concerned about alliances with their colonial rival, France, agreed not to interfere with the Indigenous people's fishing, hunting, and farming. These treaties often ended decades of outright war in which Mi'kmaq fought side by side with French soldiers and raided British ships and settlements, fighting that slowed the growth of the British colony of Nova Scotia. In return for protection of their fishing and hunting grounds, the tribes agreed not to attack British settlements already established in the region.

The treaties created a system to mediate relations between the Mi'kmaq and the British and freed Britain to focus on its colonial war with the French, which it ultimately won when France formally ceded its territory in Canada in 1763. Without the cooperation of the Mi'kmaq, Britain's victory over France in what would become Atlantic Canada would have certainly been a more costly and complicated campaign.

The treaties signed by Mi'kmaq leaders in the 1700s were notably different from those the Canadian government negotiated with western First Nations in the late nineteenth and early twentieth centuries. Those western deals, often called numbered treaties, required the surrender of First Nations lands to the federal government in exchange for compensation and the creation of separate reserves. No such provisions were part of the treaties signed with the Maliseet and Mi'kmaq, and those treaties did not involve the surrender of land.

The challenge for modern Canadian governments would become how to interpret these treaties when so little other documentation survives to clarify how far they extend. As a peace deal, the treaties between the British and the First Nations seem fairly clear-cut. But in practice, over the next two and a half centuries the fishing and hunting rights included in those deals have been frequently misunderstood or outright rejected by the courts. It wasn't until the Supreme Court's 1999 Marshall decision that the Mi'kmaq's right to participate in a moderate livelihood fishery was formally recognized in the modern era.

Part of the problem is that the federal Indian Act divided the Mi'kmaq into artificial bands, settling groups on separate reserve lands scattered around Atlantic Canada when their bloodlines often linked them to Mi'kmaq people from all over the region. Michael argues that commercial fishermen who say people from Sipekne'katik have no business fishing far from their landlocked community are misunderstanding that history.

"Our rights extend throughout the entire traditional territory," he says. "When these treaties were signed, it wasn't just signed by a community. It was nation to nation. People forget these 'bands' were just created by the Indian Act. Even in the 1930s, '40s, and '50s, they were still scooping up Native people in little tiny settlements here and there and centralizing all of them. At one time, Mi'kmaq were all over this entire province, here, there, and everywhere."

In Michael's view, the commercial lobster fishery, worth more than $1 billion to Nova Scotia, has pressured the federal government to target Indigenous fishermen on the water. Despite promises from Ottawa to resolve the dispute, in the first year and a half after Sipekne'katik tried to assert their treaty right to fish out of season, federal conservation officers seized thousands of Mi'kmaq lobster traps. Michael contends that meanwhile they're turning a blind eye to the real problem, that of non-Indigenous poaching. Fishermen on both sides of the issues accuse each other of supplying lobster into the black market through uncounted cash sales—a problem that Canada's Department of Fisheries and Oceans estimates is siphoning off hundreds of millions of dollars worth of lobster from Atlantic Canadian waters every year. The department says as much as 30 per cent of all lobster landings may be coming from poaching, an unmonitored fishery that the government says steals away tax revenue and hurts conservation efforts. Licensed commercial fishermen are required to fill out daily logbooks, reporting data such as the location they fished, the number of traps they hauled and the weight of lobster they sold, valuable information fisheries officials use to gauge the health of the stock. Poachers share none of these details.

Nat Richard, the head of Canada's Lobster Processors Association, says there are poachers in both the Indigenous and commercial fisheries. But the biggest culprits, he says, are commercial fishermen who don't want to be fully taxed on their catches—so they sell a portion of their shellfish for cash, and leave that out of their logbooks. This poses significant challenges for conservation measures, and fishermen's unions that try to portray off-the-books harvests as a uniquely Mi'kmaq issue are distorting the truth about their industry, he said.

"Anybody who knows anything in this industry knows very well that we have a very widespread and, frankly, growing problem with unreported cash sales," he told me. "I'm not suggesting to you that there's not a portion of that that's associated with some of the 'new access' fishery. But believe me, and I'm well placed to know it, this is a massive challenge right across the whole industry in Eastern Canada, including areas where indigenous fisheries are hardly even a factor."

Michael argues the fight over lobster won't be resolved until Canada updates its Fisheries Act to make room for a legal Indigenous commercial fishery. Right now, the Supreme Court has recognized that right, but it's not enshrined in any law, and that omission has led to all this conflict, he says. While many First Nations have agreed to cooperate and abide by the federal restrictions, Michael says there will always be Mi'kmaq who feel no Canadian government has the authority to restrict their ability to fish commercially.

In the absence of clear rules, federal fisheries officers are still seizing Mi'kmaq boats, traps, and other gear—sometimes just taking the equipment and not pressing any charges, Michael says. When charges are laid, Michael convinces judges to drop charges against Mi'kmaq fishermen because no section in the Fisheries Act directly regulates the treaty right to a moderate livelihood fishery.

"What we have are just a bunch of fisheries officers acting outside the law," Michael says. "If you want to regulate the Mi'kmaq people, you have to put in place an actual law. Until that time, leave them be."

The Mi'kmaq fight for their right to hunt and fish outside government regulations is hardly new, and has long been a source of immense frustration among the region's Indigenous people. A century before Sipekne'katik launched their own commercial lobster fishery, Mi'kmaq were regularly being arrested for hunting for food. Many of those who faced off against commercial fishermen in Saulnierville have direct family links to great-grandparents who fought similar battles in previous decades. Robert Sylliboy, one of the first fishermen to join the Sipekne'katik lobster fleet, is a descendant of Gabriel Sylliboy, a renowned Mi'kmaq leader from Cape Breton and the first elected Grand Chief of the regional Mi'kmaq Nation.

In 1927, Gabriel was convicted of hunting muskrats out of season, and was the first Mi'kmaq person to fight his case based on his people's 1752 treaty rights—despite being unable to read or write in English. His conviction wasn't overturned until 1996, when the Supreme Court found his legal right to hunt on his traditional territory was clear and said the justice system of his day was heavily prejudiced against him. In 2017, ninety years after his conviction, Gabriel finally received a posthumous pardon and apology from the government of Nova Scotia.

But there remain some who don't want to see Mi'kmaq compete in the commercial fishery. A month after Sipekne'katik launched their moderate livelihood fishery in 2020, Robert Sylliboy received a phone call that woke him up from a dead sleep. His forty-foot lobster boat, tied up at the Comeauville wharf in nearby Digby County, was engulfed in a suspicious fire that had started overnight. Another one of his boats was nearly rammed at sea by non-Indigenous protestors angry that he was fishing out of season. His crew's trap lines were cut, and their gear stolen or otherwise vandalized. In the small Acadian fishing town where the Mi'kmaq launched their fleet, some businesses refused to serve them. Gas stations turned them away. Marine mechanics wouldn't take their calls. Restaurants told them they weren't welcome.

"It's even kind of hard to talk about, really," Robert told *Ku'ku'kwes News*, an Indigenous news site. "It was a lot of stress, high stress every day. Constantly looking over your shoulder. You had to choose where you were going to eat. It was a very hostile environment."

Lobstering has always been a competitive business, where fishermen jockey to be the first one out on the water before the sun rises and jealously guard the secrets of their personal lobster fishing grounds. Crossing the line into another fisherman's territory can mean your gear lines get cut, your traps are emptied, or your boat is vandalized. In many rural fishing communities, and especially far out on the water, law enforcement is a long way away, so lobstermen have learned to settle disputes themselves.

But Robert and other Mi'kmaq fishermen argue that the opposition to their presence in the fishery has racial undertones. This wasn't just about lobster, they say. It's why he and others filed a formal complaint to a United Nations committee on racial discrimination, accusing the federal government of racist actions in its treatment of Mi'kmaq lobster fishermen. Ottawa insisted its officers were just upholding the law that requires lobster be harvested in season and only by licensed fishermen.

Michael McDonald says race can't be removed from the lobster spat in Nova Scotia. As a Mi'kmaq man who's spent most of his life here, he's used to being followed around stores as if he's going to steal something—even if he's coming straight from court and still wearing a suit. Every day, the Mi'kmaq are being told to know their place, he says.

"That's what we face, even to this day in Nova Scotia. We've been dealing with this our whole lives. This has been a horrible experience for some of our people, being arrested and harassed just for fishing. In their minds and in their hearts, they believe they're not doing anything wrong."

13

History Repeating

MIRAMICHI, New Brunswick – Twenty years before a wharf in Saulnierville became a flashpoint in the fight over Indigenous fishing rights in Atlantic Canada, an even more violent scene played out in New Brunswick where the mighty Miramichi River empties into the ocean.

The Miramichi River, stretching some 150 miles through the farmland and woods of northeastern New Brunswick, has long been a source of food for those who live along it. Fly-fishermen from around the world travel here to test its waters and pull salmon and bass from its depths. Ted Williams, the Hall of Fame outfielder for the Boston Red Sox, loved the fly-fishing here so much he built his own fishing lodge on its banks.

But shortly after the Marshall decision in 1999, the bounty of Miramichi Bay became a key battleground in Canada's struggle over Indigenous fishing rights. Burnt Church First Nation, named after the former Acadian settlement that was set ablaze by the British in 1758, interpreted the court ruling as permission to set lobster traps out of season. In the weeks that followed, a flotilla of angry white fishermen clashed with the Mi'kmaq fishermen, chasing them around the bay, pulling up their traps and swamping their boats. The fighting continued on land, with a wave of vigilantism, arson, and wild brawls. Someone torched a ceremonial arbour on the Burnt Church reserve, and three men were hospitalized after a pickup truck rammed into a convoy of protestors.

The Burnt Church Crisis rippled through fishing communities around Atlantic Canada. In Yarmouth, Nova Scotia, more than six hundred non-Indigenous fishermen armed with rifles and shotguns blockaded the Yarmouth harbour in an attempt to keep Indigenous people from using the wharf. The Mi'kmaq responded to the threats from white fishermen by setting up a blockade on the highway that leads to the town of Miramichi, prompting the federal government to send in armed fisheries officers. When the Mi'kmaq refused to accept a deal from Ottawa to resolve the dispute, the Department of Fisheries and Oceans cracked down aggressively, ramming boats out on the bay and making violent arrests, much of it captured on video. With people watching from shore, Mi'kmaq jumped into the water to escape. One officer was filmed beating a fisherman with a baton as he tried to climb into a boat. It's a miracle no one was killed.

"Everyone was concerned that people were going to die," recalled Herb Dhaliwal, who was Canada's fisheries minister at the time. "I had commercial fishermen saying 'we're going to bring our guns and we're going to shoot anyone fishing *our* lobster.'"

I was eighteen when the Burnt Church Crisis began. I was absorbed in my first year of university and didn't really understand what the fighting was about. It was impossible to ignore, however, as it dominated the nightly news on the TV and the front pages of nearly every major newspaper in the country. I remember a few people in my community complaining about "Natives" breaking the law up on the Miramichi, but for most non-Indigenous people, the crisis passed, the Mi'kmaq largely stopped fishing out of season, and the whole thing was quickly forgotten.

But for many Indigenous people in Canada, particularly in fishing communities, the crisis marked a major turning point in their relationship with the rest of the country. Augustine Lloyd, a hereditary chief of the

Mi'kmaq Grand Council and a member of Burnt Church First Nation, warned at the time that the crisis hadn't resolved any lingering questions around Indigenous fishing rights and was bound to flare up again.

"History will show this present injustice and it will be said that the Mi'kmaq people signed under great duress," he wrote. "Peace cannot arise out of injustice, and no certainty can result from the imposing of an unequal agreement. The Crown, and Canadians, will get no lasting benefit from these 'deals' involving the annihilation of our rights, except the despair and resentment of generations of our children and people."

Two decades later, he was proved right. A new generation of younger Mi'kmaq fishermen decided to test the federal government's resolve to enforce fishing regulations by once again launching an out-of-season, unregulated lobster fishery. For years, Indigenous people had used the Marshall decision to support their right to fish, but never on a commercial scale, and certainly not in Canada's most lucrative lobster fishing zone, in southwestern Nova Scotia. That changed in 2020, when Sipekne'katik First Nation launched their ragtag lobster fleet from Saulnierville. Members of the band argued that access to fishing is their birthright and they should be allowed to regulate their own fleet without any federal interference.

"We don't need governance from the Canadian government. We govern *ourselves*," Robert Sylliboy, one of the Sipekne'katik fishermen, told me when I was reporting for *The Globe and Mail*. The response from non-Indigenous fishermen was much the same as it had been two decades earlier. A few months after Robert joined the Mi'kmaq fishery in 2020, his boat went up in flames one night while tied up at the wharf.

But unlike Burnt Church, in the fall of 2020 the Mi'kmaq fishermen seemed to have more public support, or at least a broader understanding of treaty rights, on their side. Some restaurants in Halifax made a small but symbolic gesture against the larger commercial fishery— pulling lobster, that iconic East Coast seafood, off their menu. Instead of complaining about "law-breaking Natives," the public narrative this time shifted against the vigilantes among the white fishermen who responded so violently.

Like the Burnt Church Crisis, the Nova Scotia lobster dispute had its legal roots in the Marshall decision, a ruling that left a lot of room for interpretation—including defining what a moderate livelihood actually looks like, and how that fishery should be regulated. In Canada, the federal government negotiates fishing agreements band by band, but some, such as the Sipekne'katik First Nation, refused the concessions over control demanded in those deals. Without agreement, Ottawa left things unresolved for years.

"I don't think we should be surprised at all that there's a high degree of frustration among many of the First Nations. There's been a void on the part of the federal government," says Wayne MacKay, a law professor at Dalhousie University in Halifax. "The Department of Fisheries has not done enough to respond pre-emptively to what everyone should have known would be this kind of tension and unrest."

Jenica Atwin, a member of Parliament and a former fisheries critic for the Green Party who used to work at a First Nations education centre, told me that disputes like the ones that blew up at Burnt Church and Saulnierville will keep happening until the federal government clarifies exactly how Indigenous commercial fisheries everywhere can operate. The government, though, still insists it must manage all commercial fisheries in Canada, while many First Nations argue they have the right to regulate their fishing as an extension of self-governance.

"Without that clarity, and that leadership from the federal government, it's just going to continue. This is just going to keep happening again and again," Jenica says. "The history is there, the ruling is there, and it's laid out very plainly for anyone who's willing to take the time to learn about it."

Unlike Burnt Church, where fisheries officers rammed Mi'kmaq boats and engaged in violent arrests on wharfs and at sea, officials stood on the sidelines in Saulnierville. That left non-Indigenous fishermen feeling they had to step in—arguing that their main concern was conservation, worried that a Mi'kmaq fishery operating outside federal regulation would hurt the health of lobster stocks at a time when they were already contending with the uncertainty of climate change. They said anyone fishing in a lobster breeding ground during the moulting season risked destroying the resource for future generations.

Whereas American fishermen catch lobster in the summer, in Canada most fishermen have long respected the offseason, which runs from late May until late November, as a critical time for the crustaceans to shed their shells in order to reproduce. Non-Indigenous fishermen were angry the federal government did nothing to stop the Sipekne'katik fishermen during this protected time, so they seized and dumped hundreds of their traps in front of the local fisheries office in protest. But according to statistics provided by the federal Fisheries Department, when it comes to conservation the worst offenders are not Indigenous fishermen. Of the 2,252 charges laid by DFO between 2015 and 2019 for violations, including fishing out of season, all but a "small fraction" were connected to non-Indigenous fishing crews.

The fishing zone that comprises the waters of southwestern Nova Scotia is the most lucrative lobster ground in Canada, with catches in this one region worth half a billion dollars in 2019, the year before the Mi'kmaq launched their own fishery here. More than 1,660 lobster licences supply a local seafood industry that employs around five thousand people, and each licence in this zone allows a captain to use up to 375 traps. The Sipekne'katik First Nation's entry into this fishing zone represented just a tiny portion of the larger commercial fleet fishing these waters—initially just seven lobster licences, with a total of 350 traps. Compare that to the nearly ten thousand licensed lobster harvesters across the Atlantic region, divided among forty-five fishing zones, each with their own season, and the Mi'kmaq fishery represents a drop in the bucket.

"The amount of fishing done by all Aboriginals is insignificant compared to the non-Aboriginal fishery, especially on the South Shore of Nova Scotia," Wayne, the Dalhousie law prof, tells me. "I don't buy the conservation argument. What this is really about, I think, is strictly about protecting economic interests."

The region's commercial fishermen counter that while the lobster fishery may be lucrative, with jobs where you can earn more than $100,000 with just a few months' work, it's also expensive and difficult to break into. Many fishermen are carrying staggering amounts of debt to pay for a lobster licence and a boat, and some of them resent Indigenous fishermen who have received significant federal aid to enter the industry. Between 2000 and 2007, Ottawa spent $354 million on commercial fishing licences, fishing vessels, gear, and training for thirty-two First Nations in Atlantic Canada through negotiated agreements. The Sipekne'katik band—which now says it doesn't recognize the federal government's authority—was among the communities that received funding.

Herb Dhaliwal, the former fisheries minister, told me that those agreements and that funding were an acknowledgement that until the Marshall decision, Atlantic Canada's Indigenous people were largely excluded from the commercial fisheries. He argued that such programs have benefited Indigenous communities immensely, creating valuable fishery jobs and lifting many families out of poverty. But the trade-off is that they have to agree to federal regulation.

"They can't just be fishing however and whenever they want," Herb said. "They have a right to a commercial fishery, but it has to be done in an orderly, regulated way."

At the time of the Marshall decision, the value of the commercial fishery from Mi'kmaq and Maliseet—another First Nation in the region—was estimated at $3 million. By 2015, those same communities harvested $145 million worth of seafood, according to the DFO.

Some Indigenous groups, such as the Assembly of Nova Scotia Mi'kmaw Chiefs, the largest in the province, spent years trying to negotiate a deal for a moderate livelihood fishery with the Fisheries Department, but the options presented by the federal government infringe too much on their treaty rights, the group says.

The Mi'kmaq people will continue to fish as they always have, according to the assembly. What needs to change is how Canadians respond to that.

"Despite our rights being affirmed by the highest courts in the country, exercising these rights continues to bring frustrations, conflict, and hardships to our people," says Chief Terrence Paul, fisheries lead for the assembly. "Our communities are going fishing and we want to ensure that they don't have to be fearful of being harassed or charged."

The opposition to the Mi'kmaq fishery went beyond "being harassed." Three months after Sipekne'katik launched their self-regulated fleet, another Mi'kmaq man, this time from Pictou Landing First Nation on Nova Scotia's north shore, was shot at as he tried to stop other fishermen from cutting his traps.

Like other people in his community, Gary Denny was inspired by what he saw in Saulnierville and started putting out a few traps under his band's own self-regulated fishery. Some local fishermen objected to this unlicensed fishing and began cutting Mi'kmaq traps at night. But on a Sunday morning in December, Gary looked out his back window to see a boat circling his traps in broad daylight.

He rushed into his sixteen-foot aluminum boat to stop them, when the bigger fishing vessel turned toward him and sped up, before turning away sharply at the last moment. Gary says he was nearly swamped by a big wave when he heard the first gunshot. Then quickly came two more.

"It was just, 'Bam!' I could hear the bullet skimming across the water," he recalls. "It was the second shot that got me thinking, 'I've got kids back home.'"

None of the shots hit Gary, who quickly retreated back home and called the police. Within a day, Nova Scotia RCMP had arrested four men, but the incident sent a chill through Gary's community.

"When I went out there, I just wanted to see who it was. I never thought this would happen. At first they wanted to run me down," he says. "It was a scary moment. Gunshots? I never thought it would come to this."

Gary's fiancée, Sylvia Bernard, had been out on the water with her young son about twenty minutes before the confrontation and heard the gunshots from their balcony. She says the owner of the fishing vessel that tried to ram Gary's boat is well known in the community for his opposition to the moderate livelihood fishery. Until recently, both Indigenous and non-Indigenous fishermen in the area were often allied in their opposition to a proposed effluent pipeline from the nearby Northern Pulp mill, which they say risked contaminating the waterway they shared. But now, the Mi'kmaq say that relationship has been replaced by chronic vandalism, despite the small scale of the band's fishery.

"Everyone's traps just kept getting cut and cut. But we could never catch the guy," Sylvia says. "It's not like we're cleaning out the ocean. We only put in about twelve traps each."

As she tells me this, she can see two grey federal fisheries boats patrolling the water near the reserve, brought in to ease tensions. But anxiety in the community remains high, and some are wondering if they should put guns in their boats to protect themselves.

"It's tense here," says Edgar Denny, a sixty-year-old fisherman who lives in Pictou Landing. "If they're shooting at us, does that mean I should arm myself and start shooting back? Retaliation doesn't get you anywhere."

Gary vowed to be back on the water next spring.

"People don't like what we're doing, I understand that. But we're not going to give up," he says. "This is about more than just lobster. This is history being made. I want my kids to be able to fish without going through what we're going through."

Older Mi'kmaq say the reaction to Indigenous people catching lobster is purely emotional, because that fishery has brought so much money into non-Indigenous fishing communities. Dave McDonald, a Sipekne'katik Elder who is approaching seventy, has been harvesting halibut, scallops, gaspereaux, moose, and deer from Nova Scotia's waters and woods without a licence since 1972. But none of that provokes anger in Nova Scotia quite like lobster, he tells me as he sits in his minivan by the Saulnierville wharf.

"As soon as we start making some money for ourselves, they want to put us down," he says. "They've got a privilege to fish, but we've got a right to fish."

All along Nova Scotia's southwestern coast, where generations have earned a living from the sea, few issues provoke as much fury as Ottawa's management of the fishery. Fishermen here complain that the government has allowed the Indigenous fishery to expand out of control, risking the livelihood of thousands. Some reserves' insistence that the federal government can't regulate them, force them to fish within certain seasons, or impose licences or any other restrictions on their catch has non-Indigenous fishermen deeply worried.

"Where does it end? Where's the limit?" asks Sue Beaton, an Antigonish fisherman who learned how to fish from her father. "Even small-scale changes to the fishery can be detrimental to the species. You can't manage a fishery this way, because the lobsters don't care who's catching them. A lot of people in our industry are pretty pissed this has been allowed to happen."

Sue tells me she paid $350,000 for her lobster licence twenty years ago, and figures that with loan interest it has cost her closer to $650,000. Like a lot of fishermen, selling that coveted commercial licence is her retirement plan. But she fears that the dispute with the Mi'kmaq is hurting its value.

"I think all of us are a little worried about how big this is going to grow, and whether or not we're going to be displaced," she says. "Which is frightening when you have such a big investment. You can't expect a young guy to spend half a million dollars on a lobster licence if you have this question hanging over everything."

On both sides, they point an angry finger at the government in Ottawa. That anger spilled over into the 2021 federal election, costing the federal fisheries minister, Bernadette Jordan, her job when voters in her coastal riding chose a Conservative candidate who took an anti–Indigenous fishery stance. Erin Crandall, a political science professor at Acadia University in Wolfville, Nova Scotia, says a lot of political capital has been spent over the years trying to fix the divide between Mi'kmaq and white fishermen. But when it comes to actually finding a lasting resolution to the dispute over the growing Indigenous fishery, one that all First Nation groups agree to, that's proved to be a challenge.

"In terms of political issues in the region, it's one of the biggest and most important," she says. "It's a question of how do we reimagine the relationship between First Nations and the Canadian state. But there's no easy answer."

There's no question that the lobster dispute is fracturing fishing communities. Some fuel suppliers, marine mechanics, and rope companies refuse to supply the Mi'kmaq fleet. But it's also creating bitter rifts between—and in some cases, within—families.

Christina Ward, who has mixed Acadian and Mi'kmaq heritage, says her decision to work with the Sipekne'katik fishermen caused some family members to complain that she's stealing other people's livelihoods. She's worried her teenaged children will be targeted at school because their mother works for the Indigenous fishery. There have already been fistfights between the children of fishermen on both sides, she says.

"I'm worried about my kids going to school, because I know there's going to be trouble," she tells me, squinting at the RCMP truck patrolling the Saulnierville wharf. "I had to teach them how to defend themselves."

But the money is decent—up to $2,000 a day for a good catch. The only other way she could make that kind of cash, she tells me, is selling drugs.

For others, like Dave McDonald, the Mi'kmaq Elder, it's an issue of equality. Nova Scotia's Mi'kmaq remain among the poorest people in the province. Commercial fishermen, he contends, drive bigger trucks, have better boats, and enjoy more comfortable lifestyles.

To better understand that disparity, I visited the main assembly shop at Yarmouth Boatworks, where a haze of fibreglass dust was mingling in the air with the sound of classic rock blasting from a stereo. Here, workers were building $1.2-million state-of-the-art vessels, complete with kitchens, sleeping quarters, flat-screen TVs, and showers, designed to venture far out into the ocean and carry thousands of pounds of lobster back to land. They're a far cry from the secondhand thirty-five-foot boats most Mi'kmaq use.

Owner Steve Gee tells me demand for these giant new boats has slowed down as fishermen begin to second-guess investing in the fishery. But still, lobstering remains good money, he explains, and that's the main reason he can't get enough skilled workers to run his shop at twenty-eight dollars an hour. Not when they can earn six figures as a deckhand on one of the thousands of fishing boats in Nova Scotia.

"I need twenty-three workers, but three showed up today," he says. "Our latest job posting had three hundred applicants, and not one of them was in Canada."

The Mi'kmaq, meanwhile, say they're tired of being excluded from the riches of lobstering, and they're finally taking their share. They don't intend to stop, regardless of what the law says, Elder Dave McDonald tells me.

"Look at the houses they've got, then come onto the reserve and look around," he says. "Just because you spent a million dollars to fish doesn't give you a right to fish, but that's what they think."

14

A Vow to Fight

DELAPS COVE, Nova Scotia – Far across the water from Brad Small's house, on the Nova Scotia side of the bay a boatbuilder named Colin Sproul is speaking passionately about the future of lobstering. His voice is deep and a little hoarse, like someone who's been shouting into the wind for years. Colin lives in Delaps Cove, a small dimple in the long Bay of Fundy shoreline, the kind of place where people still gather to watch new boats being launched into the sea. Boatbuilding is a relatively new venture for him, but lobstering has always been the family business. His son is the sixth generation of Sprouls to make a living at it.

Colin, however, has a new job, one that consumes most of his energy. As president of a group called the Unified Fisheries Conservation Alliance, or UFCA, he's at the centre of what many Canadian fishermen see as an existential fight for the survival of their industry. A long-time fisheries union leader, Colin became heavily involved in the Indigenous fishery issue when it started making national news in the fall of 2020, erupting in riots, fires, and violence on the water.

"I was dragged into it kicking and screaming," he says. "But it's everything I am, you know what I mean? I grew up as a fifth-generation fisherman, it's all my family has ever known. And I think that at some point, people in Ottawa need to understand that we don't have anywhere else to go, this is all we have. It's all we know."

The UFCA, an alliance of fisheries associations from around the region, was formed to represent fishermen in court cases and to speak with one voice when lobbying on the issue of the treaty fisheries. Colin says he decided something needed to be done when an older fisherman he'd known for years came to him, worried about what the expanding First Nations fishery would mean for the next generation of non-Indigenous fishermen.

"He basically came to me in tears, saying, 'I don't know what to do here. My son is just about to graduate. Should I be pushing him to go to university? I wanted so bad for him to follow behind me and his grandfather into the fishery. But I don't know if I'm sending him down a dead-end road, like what am I supposed to do?' And it was just the final straw for me, I knew that I had to stand up for him. I had to stand up for voiceless people," Colin says.

To say fishing and the sea is a part of life along this shore is an understatement. In many small communities, it is *all* they have. This coastline has given birth to, and taken the lives of, countless fishermen, shipbuilders, and seafarers since settlement began here. One of the most famous among them, Joshua Slocum, was the first person to sail single-handedly around the world, in 1898. He died in 1909 while sailing from Massachusetts to the Caribbean, taken by the ocean that he loved so much.

The sea has always provided here, but people are very worried about the future in a way they weren't in previous generations. People who earn a living in the fisheries feel they're under attack from multiple sides: by climate change, by new whale regulations and offshore wind development, by corporate concentration among the seafood processors they must sell to, and by an Indigenous fishery that grows every year.

Colin contends that most people in the fishery support the Indigenous right to earn a living from fishing. But he gets angry when he sees that right to fish being leased out to non-Indigenous fishermen, which is already happening on boats around Atlantic Canada. What happens, he asks, if a First Nation wants to sell its fishing rights to a foreign company desperate to get access to Canada's lobster harvest? That's not, he believes, what the Supreme Court intended in its Marshall decision.

"It's a participatory right," he says. "It's not an absolute right, that every Mi'kmaq person should be delivered a moderate livelihood if they participate in the fishery. It's the right to go try to make one like everybody else."

His worry that rights-based access to the fishery can wind up in the hands of a foreign corporate interest is not unfounded. New Zealand's indigenous Māori signed a treaty with the British Crown in 1840 that guaranteed "undisturbed possession" of the fisheries until they chose to dispose of them. A long-standing claim was settled in 1992, allowing the Māori to purchase 50 per cent of a company that held a major portion of the country's fisheries quota. Then, in 2004, the Māori obtained 20 per cent of all aquaculture around New Zealand's coasts and harbours through the Māori Commercial Aquaculture Claims Settlement Act.

Since then, Moana New Zealand, the largest Māori-owned fisheries company in the country, sold a 50 per cent stake in the firm to a Japanese seafood giant, forming Sealord. The partially foreign-owned company now uses its exclusive access to the fisheries quota to harvest from the waters around New Zealand and sells almost all of that catch to buyers in Japan.

Colin says nothing prevents this from happening in Canada. He points to the Clearwater deal, which made a handful of Mi'kmaq bands 50 per cent owners of the largest lobster company in the country with exclusive access to the deep-water offshore harvest. It was celebrated as a progressive step forward, yet it doesn't benefit all Mi'kmaq First

Nations, Colin argues. And the people it benefits the most are the investors in the private equity firm in Vancouver.

"There is one simple solution to what's been an incredibly complex, devastating problem for the Maritimes. And it's to stop non-Indigenous participation in individual rights-based fisheries, which is what anything delivered through the Marshall response is—it's the right of an individual Mi'kmaq person to participate to get the chance to earn a moderate livelihood," he says.

The issue of treaty fisheries has been simmering in Canada for several decades. Now it can no longer be ignored, Colin says. Feeling pushed into a corner, commercial fishermen aren't going to back down this time. His group says it can't wait for the federal government to take action, and has begun investigating fishermen and buyers who it suspects of fuelling the black market. In August 2024, the Unified Fisheries Conservation Association launched a lawsuit against a lobster pound in Nova Scotia's Shelburne County, alleging it was buying illegally-caught lobster.

"We're not going away," he warns. "We're not going peacefully into the night. We're not going to see our community's future transferred to some other community in the spirit of reconciliation. I believe in the spirit of reconciliation, but I believe it's a price that all Canadians deserve to bear equally. And right now, only some Canadians are bearing any price for it, and especially rural communities, and it's just, it's wrong. And two wrongs don't make a right. Colonialism was clearly wrong, what it did to Indigenous people. But destroying those precious fishing communities as retribution doesn't make anything better. And that's what I'm standing up for."

15

The Lobster Cartel

METEGHAN, Nova Scotia – When the Riverside Lobster plant in Meteghan, a rural Acadian fishing village in the heart of Nova Scotia's French Shore, closed in the fall of 2023, ending hundreds of jobs, its owners gave an unusual reason. They were running out of lobster, they said. November in southwestern Nova Scotia is typically the peak time for the Canadian lobster harvest. And yet Riverside's owner, a Quebec-based private equity group called Champlain Financial Corporation, was blaming a pronounced lack of supply for the closure.

"The lobster processing industry in Atlantic Canada is continuing to see an unprecedented situation, with not enough lobsters to sustain current processing capacity," company spokesperson Rachelle Gagnon explained. Indeed, anecdotal reports from around the province suggested catches were down notably in some areas in the critical early weeks of the season. Heather Mulock, executive director of the Coldwater Lobster Association, which represents fishermen in southwestern Nova Scotia, said the fall of 2023 saw one of the worst catches in the region since the 1990s. Some local fishermen reported catches were half what they were the year before.

Some immediately blamed the Mi'kmaq fishermen harvesting from nearby St. Marys Bay during the summer months, when the commercial fishery is closed. Indigenous harvesters pushed back against the suggestion that they're responsible for the declining lobster population, pointing out that 94 per cent of Nova Scotia lobsters are still harvested by the non-Indigenous commercial industry. Meteghan is in the municipality of Clare, which lies at the heart of the growing Indigenous fishery in Nova Scotia, where some fishermen claim that the out-of-season lobster harvest has grown so large and reportedly so embedded with organized crime that some federal fisheries officers refused to do enforcement work there.

But there was another issue behind the decision to close the Meteghan plant that Champlain didn't want to discuss. Increased consolidation among seafood processors, fuelled by a buying spree in the industry over the preceding decade, means it's easier for companies to close down facilities to trim costs and force fishermen to accept lower prices for their catches.

Between 2017 and 2021, Champlain bought eight processing plants in small rural communities across New Brunswick and Nova Scotia. When one of those plants burned down in 2020, Champlain didn't bother to rebuild, throwing 150 people out of work. The decision to close Riverside Lobster was made *before* the poor 2023 season even began, and reduced catches only provided convenient cover.

Colin says, "I warned the federal government that this would happen if we allowed a handful of corporations to take over all aspects of the lobster supply chain and vertically integrate it, and they looked at me like I had two heads. Now we're seeing massive consolidation, and this is the consequence."

Increasingly, control over the lobster supply in Canada's three Maritime provinces lies in the hands of just two companies, Vancouver private equity firm Premium Brands and Montreal's Champlain Financial, which is owned by an affiliated U.S. private equity fund. In 2021, Premium bought Clearwater, the world's largest lobster company, and in 2022 tried to acquire Champlain, a deal that would have created a seafood behemoth if it weren't for crashing crab prices that summer.

In Newfoundland, a Dutch Crown corporation, Royal Greenland, owns more than half the fish buying and processing market. Keith Sullivan, then head of the province's largest fisheries union, says the company's end game is to eliminate competition and force fishermen to accept lower prices.

"Royal Greenland's actions are a blatant and egregious misuse of a processing license—something that is a privilege and must be used to the benefit of the people of our province," he said in a statement posted on the union's website. When I approached the company for a response, they declined to comment on these allegations.

Colin says this consolidation across the industry has the effect of pushing down prices that lobstermen get for their catches. When processors like Riverside close, it removes options for local seafood buyers to sell lower-quality lobster that can't be sold into the live market. At the same time, the seafood companies are buying up packers, shippers, and bait suppliers, giving them more control over every step of the lobster business.

"This allows the processors who are still existing to offer a lower price," says Colin. "I don't think there's anybody left they can sell their lobster to who aren't owned by Champlain or Premium. Then it doesn't take much for two guys to put their heads together and fix the price." Champlain and Royal Greenland declined to comment on these complaints from Colin, Keith, and others for this book. Premium Brands did not respond to a request for comment.

While companies like Premium actively increase their ability to push for lower wholesale prices, fishermen have limited options to push back. They can refuse to fish, which happens occasionally. But most need the income and can't afford to keep their boats tied up.

In Maine, lobstermen are prohibited by law from going on strike, but have used collective action in the past to push for better prices.

Most famously, after four thousand fishermen tied up their boats in 1956 and 1957 in an effort to get a minimum price of thirty-five cents per pound of lobster, the U.S. Department of Justice indicted the Maine Lobstermen's Association under the Sherman Antitrust Act. Passed in 1890, the Sherman Act prevents companies from maintaining a monopoly and from controlling prices or supply of a product or service. It's best known for helping break up Standard Oil and the American Tobacco Company.

"What 'evil' had the fishermen done?" the fishermen's lawyer, Alan Grosman, told the jury, according to an account in the *Cape Cod Times*. "They had resisted a price imposed on them, on which they could not make a living, and so tied up their boats and put their lobsters in a bank—God's bank, the bank of the Atlantic Ocean."

The government eventually backed down, knowing that some Portland lobster dealers had already been indicted for price-fixing. The judge even told the jury that although the dealers were clearly colluding to push down prices, the law prevented the fishermen from responding. In the end, all fines against the lobstermen were withdrawn.

Martin Mallet, head of the Maritime Fishermen's Union, which represents over 1,300 independent inshore owner-operator fishermen—those who work on their own boats—in New Brunswick and Nova Scotia, said the modern version of consolidation is a deeply concerning trend. Handing control of the lobster industry to a small group of private equity firms takes away fishermen's right to choose where they sell their catch and reduces competition, allowing processors to squeeze fishermen on price.

"In recent years, mom-and-pop and family-run and community-based fishing and processing enterprises have been on the international menu to be bought and agglomerated by large corporate interests owned by foreign nationals or out-of-province investors. Whereas locally run enterprises would reinvest most of their business revenues within their community and province of origin, large-scale corporations are interested in profits for shareholders, not sustainable rural communities that depend on a local fishery's resources," he told the House of Commons Standing Committee on Fisheries and Oceans in 2021.

Investors have flooded into a lobster industry that until recently was still marked by independent companies that came and went through boom-and-bust cycles. These new equity funds are attracted by projections that prices for wild-caught seafood will continue to climb globally as catches decline.

"Times are changing," Martin says, "and this is a very high-value industry now. And it's a limited resource as well, the fisheries resources are limited. So they're expected to increase in value in the upcoming decades. Their intent is to control the buyers and protesters in these provinces. And once they do, they typically force whatever shore price that they want to fishermen."

What people don't want, he says, is a return to the era of the 1940s through the 1960s, when many fishermen were basically indentured servants to large international processors. These vertically integrated companies exerted significant control over the fishery, from fishing to processing to marketing. They owned large processing plants and huge fleets of trawlers, and were widely blamed for the collapse of the groundfish sector on the east coast.

Today, because of the extremely high value of lobster licences in Canada, some fishermen are turning to seafood processors instead of banks to finance the cost of entering the industry. And that, Martin says, leaves them indebted to the people they sell their catch to, which further drives prices down. Fishermen also complain that corporations are cornering the market on bait, a critical supply all lobstermen need to fill their traps, and are using that as leverage to demand loyalty. In decades past, lobstermen would buy their bait directly from herring fishermen, but as that fishery has disappeared, they've come to rely on lobster buyers to provide it.

Canada's Fisheries Act is supposed to protect the inshore fishery from potential takeover by large corporate interests by requiring that harvesting be done exclusively by independent owner-operators. But in reality, Martin says, many fishermen are increasingly indentured to seafood companies.

"You have instances where some of these processors are lending the money, but also with the lending of the money comes agreements around buying and controlling. So we are seeing a huge loophole to the Fisheries Act that's supposed to protect the owner-operators. By having this loophole in the back end, they're having more and more fishermen connected to specific processors. And this is probably the biggest change that I've seen in the last twenty years."

Colin calls the corporate concentration of ownership "the lobster cartel" and says the power this group exerts is not unlike that wielded by the Hells Angels, the notorious biker gang that controls the wholesale price for cocaine in Canada. But unlike the bikers, who use violence and intimidation, the lobster cartel is perfectly legal, using their buying power and monopoly to control fishermen, he says.

"It's probably the worst thing that's happened to lobster fishermen in Canada in decades and nobody's rung the alarm bells," he says.

While city folk may have a romantic picture of lobstermen selling their catch directly off the boat, for most it's not realistic to sell their highly perishable shellfish that way, or not in any significant volumes. They need to work with lobster wholesalers, whether they like the price or not.

"It's really hard to sell ten thousand pounds of crap-quality lobsters on Christmas Eve in a snowstorm," Colin says. "It's easy to sell top-quality lobsters in July at the end of our driveway with a million tourists driving by."

The counterpoint is that wholesalers take on risks that most lobster fishermen won't. When they buy "shoreside" lobster—meaning directly from fishermen—they ultimately have to unload them regardless of what processors and shippers are willing to pay. And these arrangements usually mean buyers have to buy *all* the lobsters from fishermen they deal with, including one-clawed, damaged, or barnacle-laden lobsters that are destined to be processed because consumers won't want them whole. As the price goes up and down, the wholesalers have to absorb any loss. It's not unheard-of for dealers to sell lobster at a loss when they've run out of storage space and the market is saturated.

"If you're a lobster buyer that buys from the shore side, no matter what size you are," Colin says, "eventually you'll be in a position with your business where you've got a whole bunch of lobsters held in your holding facility. And they've got decreasing blood protein levels, or some other indicators that show that they're not going to make it to live market. And you've got these huge risks associated with being a buyer, so that you're going to have to do something to unload $2 or $3 million worth of product sitting in your pen. Fishermen can't really operate like that. You've got to have a whole separate side of your business to do that, with all the wins and losses."

Geoff Irvine, executive director of the Lobster Council of Canada, a marketing board for the industry, acknowledges that the pricing model for lobster can at times be frustrating for fishermen. But he says they need to remember that consumers have choices, and will buy other proteins if the price of lobster gets too high. He points to 2021, when a surge in global demand for lobster pushed the price through the roof, forcing some restaurants to take it off the menu.

"If we push the prices up too high," he says, "consumers won't buy us at retail. And we come off menus, because shrimp is cheaper or pork is cheaper. So that's a huge concern, a very serious concern. And there's only so much lobster, and so, yes, the shore pricing system is frustrating. It sometimes doesn't make any sense. But it's all we have."

Colin contends that the big lobster companies are trying to chip away at the price gains lobstermen have enjoyed over the past two decades, fuelled largely by the creation of new markets for their catch. Those new markets, in such places as China, Vietnam, and South Korea, have opened up new avenues to ship live lobster—which fetches a premium price over the processed variety—that didn't exist not that long ago.

"All these new markets have exploded," Colin says. "When I was a kid, we had almost no market in Asia. Eighty per cent of our lobster went to the U.S., and 20 per cent was in the Low Countries, in Europe. And now we've got 40 per cent of our market in Asia, 40 per cent in the U.S., and 20 per cent in Europe. And that market in Asia, where they appreciate quality, and will pay for it, has led to an increase in lobster prices. Everything has changed now. And I don't know how anyone can look at what's happening in the fishery today and not be alarmed by it."

The "cartel" enforces loyalty among buyers by punishing those who pay more than the established price for lobster, Colin says. In his view processors will blacklist anyone who breaks rank and refuse to buy from them. And the big exporters will go after anyone who tries to sell those lobsters in other markets, wherever they may be, he says. The big seafood companies, of course, deny all of these allegations and say the market sets the price for lobster.

"They'll undercut you at a loss at your market. They'll find out where you're selling your lobsters in Boston, or Shanghai, and they'll offer them to that customer cheaper, even if it means a loss for them. They'll do that over and over again until you're out of business," he said.

Nat Richard, the head of the Lobster Processors Association that represents many of the companies Colin is referring to, has heard plenty of fishermen complain about this before. He thinks the idea of a lobster cartel that fixes prices is ridiculous. There has been consolidation in the industry, in both Canada and the U.S., but there remains enough competing buyers that smaller, independent processors often gripe that the largest companies can afford to pay more for their lobster, he said.

"This notion that we have a behemoth that's going to control lobster, I just don't buy that. We have multiple groups. There has been consolidation, certainly, but this is still a very, very competitive industry," he says.

Lobster is a tough business, even for the biggest players, he said. He points out that Clearwater, one of largest seafood companies in the world, had to close two of its lobster facilities in Nova Scotia in February 2025, in the face of stiff price competition in the live lobster export business. The corporation said it wanted to focus on processing and freezing shellfish and other seafood on ships at sea instead.

There are many things that can affect the price fishermen get at the wharf for their lobster, Nat says. That includes changes in the amount of lobster caught at certain times of year, and volatility in an export market that relies so heavily on overseas demand. There also remain too many fishermen and buyers willing to deal in unreported cash sales, he says, which can have a detrimental effect on pricing.

"It's always bizarre to me to hear people argue, 'Oh my God, these groups are going to drive down prices and exert control on price.' I hear the opposite view from many other plants on almost a daily basis," Nat says. "I really do think that it all comes down to fundamental supply and demand forces. And we see that all the time, every year, every season. That's what drives the price on the shore."

PART THREE

A GLOBAL FOOD

16

———

Chinese Lobster

QINGDAO, China – The ocean's buffet is on offer in Qingdao. Peruvian eel. Farmed shrimp from Ecuador. Brown crab plucked from the Irish coast. Turkish red mullet. Silvery ribbonfish from India. Ungava prawns caught by the Inuit of northern Quebec. Cod filleted at sea by Russia's floating factories. Caviar from Singapore. Slabs of orange salmon raised in pens in Norway. On and on, an endless array of seafood from around the world is being showcased.

The China Fisheries and Seafood Expo, one of the largest trade shows of its kind in the world, is a dizzying display of the bounty of the sea, and of the industry's determination to sell it all to China, the largest seafood market in the world. Held over three days in the port city of Qingdao, on China's southeast coast, the expo sees more than thirty thousand visitors each year, many of them buyers for Chinese importers, distributors, restaurants, and grocers.

Spend a few days walking around this massive show, spread over eighteen cavernous convention halls, and it's easy to see how important China is to those who sell seafood around the world. With more than a billion people and a rapidly growing middle class, China consumes more than sixty-five million tonnes of seafood every year, about 45 per cent of the world's total. Americans, by comparison, eat just seven million tonnes.

"It's crazy," says Alex Puig Coll, a veteran seafood salesman from Barcelona, watching the crowd move around the trade show scanning for free samples and brochures. "The dimensions of this market are so big. I mean, it's *endless*. China could take all the lobster in the world, and we still wouldn't have enough to meet the demand."

Alex is director of international sales for the Lobster Trap, a company in Bourne, Maine, that has been exporting lobster since 1972. For companies like his, the European markets that were the first to buy North American lobster are declining as the price of lobster keeps creeping upward. Increasingly, Asia, and particularly China, is their focus. And they see huge opportunity for growth that is only just beginning.

It's a market that hardly existed little more than two decades ago. The explosion in China's appetite for North American lobster has mirrored the country's desire to be consumers of everything the globe has to offer to those with money to spend, from smartphones to high fashion to electric cars.

"Twenty years ago they were all driving bicycles. Those who had cars were just learning to drive them. It was a complete shit show in the roads," Alex says. "People don't understand that China hasn't just caught up with us, they've passed us. The Chinese don't want to just eat rice anymore. They still eat rice, but they also want hamburgers, Starbucks, everything the rest of the world eats."

The convention centre he's sitting in, which would dwarf most international airport terminals, is a towering monument to Chinese determination and economic power. A few years ago, this site was rice paddies. Today, it's 144,000 square feet of convention space that feels like a small city. Outside, luxury vehicles whip by on the highway as construction cranes work away at a national basketball training facility across the road. The entrance to a brand-new subway line that links the convention centre to the city of eight million, spotless and graffiti-free, is less than a block away. A black armoured police vehicle is parked out front, with officers standing guard behind riot shields, serving as a reminder that you're still in China. The crowd pay the show of force no mind, however, as street vendors hawk dumplings and noodles to the near-constant flow of visitors.

Until the late 1990s, most people in mainland China had never seen an American lobster up close, let alone tasted one. In those years, seafood imports were highly taxed by the Chinese government, and only state-owned companies could bring in shellfish from beyond China's waters—and only at a rate that matched their domestic catches. China began to loosen the rules in the early 2000s, after it joined the World Trade Organization, but most lobster coming into the country was still being smuggled in through the "grey market" in Hong Kong. Shipments of Australian rock lobster began to trickle in, along with abalone and phallic-looking geoduck clams, but in those early years few Chinese could afford these foreign shellfish.

It wasn't until American exporters began shipping live lobster from Boston international airport in the late 2000s that North American lobster began showing up in Chinese seafood markets in large volumes. The Chinese buyers called this new shellfish, with its big, meaty claws, "Boston lobster" because that was the port stamp it arrived with, a name that still sticks for many Asian consumers today. It was about half the price of the Australian lobster being sold in the country, and quickly became seen as a more affordable and more abundant alternative.

China's economy grew rapidly, creating a massive middle class with disposable incomes, and they started seeking out new ways to show off their wealth. In a culture where face, or your public image—known as mianzi in Mandarin—is still highly coveted, few things impress guests more than a table piled high with lobsters imported from the other side of the world. That's a big reason why *Homarus americanus* has become a staple of Chinese weddings, buffet restaurants, and five-star hotels in the past two decades.

"For the Chinese, the more you spend on dinner, the greater face you have. So that's why they started importing foods," explains Peter Redmayne, president of Sea Fare Expositions, a Seattle-based company that runs the China Fisheries and Seafood Expo.

As China opened up, a new market for North American lobster was created almost overnight. From 2010 to 2017, Canadian lobster exports to China rose from around $1 million to almost $173 million. Within another five years, live lobster exports had grown to $455 million—nearly surpassing the value of lobster sold to the U.S., historically the largest buyer of Canadian lobster. By 2020, Canada and the U.S. were exporting more than $2 billion in processed and live lobster to China, the vast majority of it live.

"They started serving it in high-end five-star restaurants. But then they found it was pretty easy to develop new markets because the demand was there. And it was cheaper than Australian live lobster, so it just started to take off. The exports just went crazy, and that demand increased greatly right up until the pandemic," says Peter, who was editor and publisher of a seafood trade magazine in the early 1980s before he started organizing seafood trade shows in California.

COVID-19 interrupted trade around the globe, and in China it sparked a recession that has slowed spending. But many seafood importers, such as Alex, are betting on Chinese demand rebounding, and they see the potential for lobster to be much more than a special-occasion food there. One day soon, he says, we may see an army of lobster roll shops spread out across China, as they have already across the U.S. and increasingly in Europe and Japan. Value-added lobster products, processed in Canada and shipped frozen to China, are beginning to be sold. Indeed, companies like Clearwater, the largest seafood company in Canada, are already selling frozen lobster tails in China. Royal Greenland, the Dutch company that has cornered the lobster processing market in Newfoundland, is doing the same. Alex tells me that an Italian company is exploring setting up a processing facility in China to make lobster sauce, a popular garnish among Chinese diners, from all the waste shells and heads that are currently unused by the industry.

"I think that's the next wave," Alex says, pointing to a nearby booth where frozen seafood entrees are on display.

Everyone at the trade show connected to the lobster industry seems to be convinced the market's ceiling is a long way off. Global exports of lobster increased again during the first half of 2023, by more than 10 per cent compared with the same period in 2022. Canada, the largest lobster exporter by far, sold nearly fifty-one thousand tonnes to buyers in other countries, an increase of nearly 8 per cent. U.S. exporters sold about half that much in the same period, and watched their total lobster exports decline, largely because the catch is shrinking in American waters. More than a quarter of all lobster exported in the world now goes to China, and most of that comes from Canada.

When U.S. President Donald Trump began his second term in 2025 by threatening 25 per cent tariffs on most Canadian goods, many in the Canadian lobster industry argued Canada should be doing more to grow its market in China, already the country's second largest trade partner, as a buffer to the whims of the White House. The Chinese government, for its part, said it was open to discussing greater trade ties with Canada. But the lobster market can be a volatile and unpredictable place. Within weeks of saying it wanted more trade with Canada, China—which bought $1.3 billion in Canadian seafood in 2024—announced its own plans to slap 25 per cent tariffs on a long list of Canadian products like lobster, snow crab, and shrimp.

That was the same levy they used against American lobster in 2018, which devastated U.S. exports to the country. Seafood exporters who lived through that experience know it can take years to build international trade relationships, but they can be dismantled in a matter of days. Canadian fishermen worried a drawn-out trade war with China could be a nightmare scenario for the industry, considering China buys 40 per cent of their live lobster.

Between warming oceans, extreme weather, and changing fishing patterns, the lobster fishery is used to unpredictability. But geopolitical conflicts have added a new layer of challenges. The National Fisheries Institute, the organization that represents the American seafood industry, has lobbied hard in Washington to keep Canadian lobster off Trump's hit list, arguing Americans don't want to pay even more for the shellfish. About three-quarters of all frozen or processed lobster that leaves Canada is sold in the U.S.

While Trump eventually backed down on his plans for broad tariffs on Canadian foods, including lobster, the trade wars have left fishermen on both sides of the border feeling uneasy about the future and worried that the old rules of lobstering are increasingly unreliable. In response to Trump's tariff threats, some Canadian politicians were hastily dispatched on global trade missions with seafood companies in tow, trying to develop new markets outside of America. Nova Scotia, which exports $2.5 billion worth of seafood, said it couldn't afford to sit idle in a protracted trade war with the U.S.

"It has never been more important to showcase our premium quality seafood on the world stage," Nova Scotia Fisheries and Aquaculture Minister Kent Smith told reporters in January 2025. "With the continued uncertainty from the United States, it's more important than ever that we ramp up our efforts to help Nova Scotian companies expand into new markets."

Finding new markets for lobster as tariffs come and go is easier said than done, of course. The Canadian and American seafood industry has been interwoven for centuries, and that can't just be switched off. Maine, for example, sent 40 per cent of its lobster to Canada for processing in 2024, and most of that was trucked back across the border as claw and knuckle meat, frozen tails and other products.

China is a massive buyer of live lobster, and a trade war with such an important overseas customer has far-reaching impacts for exporters in Canada and the U.S., as recent history has shown. But a tariff fight between Canada and the U.S., the two countries that actually catch those lobsters? The industry warned it would be deeply painful for fishermen, processors, and even consumers, who ultimately pay for the increased costs brought on by tariffs.

"I mean, geography matters," says Nat Richard, the processing industry spokesman in Canada. "And that goes back centuries. We've been processing each other's lobster for a long time. And so this notion that we're just going to reorient and pivot to other markets? Well, there's just no way another market can absorb these volumes. This is a big challenge for the industry."

As China kept buying more and more lobster, rising prices became the new norm. While prices fluctuate with the fishing seasons, and are affected by short-term gluts, pandemics, and global recessions, the money seafood companies pay fishermen for their catch has been increasing steadily. In 1998, lobstermen in Maine were getting $2.92 a pound for their harvests. In 2021, they were getting $6.71 per pound. That increase, driven by global demand, inflates the costs to processors and exporters, who ultimately pass it on to consumers. As the supply of lobster tightens as landings decline in the U.S., this trend is only expected to continue.

"When the market accepts a certain price, that becomes the new floor," Alex explained.

Logistical improvements have also made it possible for live seafood to reach the interior of China much more efficiently now, opening up access to new consumers in cities that used to be out of reach for importers. When Alex first started selling seafood in China, distributors would place their orders by fax machine and pick up live lobsters from the airport in Beijing in unrefrigerated trucks. In midsummer, temperatures there often reach above 30 Celsius—a death sentence for a lobster out of chilled water. Mortality was a common problem. It was the "Wild West," Alex says. Today, exporters and importers understand the importance of refrigeration when it comes to a perishable food like live lobster.

The whole world wants to sell lobster to China, it seems. North American lobster dominates the market, but it's still competing with Caribbean spiny lobster from Cuba, European lobster from Ireland and the U.K., and rock lobster from Australia and New Zealand, all of which are more expensive. American lobster is considered the more affordable option, but it's still costly enough to be a luxury item. Yet no matter what species is being sold, there remain challenges for companies importing the pricy shellfish here, and there have been costly lessons adapting to the quirks of the Chinese market.

"China is still a frustrating market sometimes," Peter says. "It's very easy to sell there before their big Chinese New Year, when people will pay so much more, and the demand can be very high. And then it suddenly falls off, and the price usually goes *way* down. So if you get stuck with high-price inventory it can be very painful. But I mean, you can't deny the size of the market."

If one thing was pulsating through the trade show in Qingdao, it was optimism. After a three-year hiatus because of the pandemic, the China Fisheries and Seafood Expo showed that Chinese importers are hungry to expand their offerings. Even as China's economy slowed to a more earthly pace of growth, plenty of industry watchers here expressed confidence that the market has a lot of room to grow.

It's hard to imagine how that's possible, but China's appetite for the world's seafood seems to know no bounds. Between 2018 and 2022, the country imported between sixty-two million and seventy-two million pounds of lobster annually. Most of the growth was in imports of live Canadian lobster, a market that grew from more than $172 million in 2017 to nearly half a billion dollars in value five years later—and that's with a pandemic that regularly shut down the country's biggest seafood importers. That kind of exploding demand upended the old rules and economics of the lobster industry, and increasingly forced companies to think globally instead of regionally.

A century ago, Qingdao was little more than a poor fishing village and a strategic naval station for the German empire. Those colonial years left the port with broad streets, European-style infrastructure, and the Germania Brewery, which later became the world-famous Tsingtao Brewery. The cobblestoned streets of the city's old town run uphill to a colonial-era cathedral, but at night, the crowds here gather at shrines to Western food—namely, KFC and McDonald's, whose signs light up the streets a few blocks away.

Today, Qingdao is an affluent coastal city of nearly eight million people, with an important network of seafood importers and a downtown filled with an ever-growing number of glass skyscrapers huddled along its sandy shoreline. At night, the city skyline lights up in a choreographed light show, in the way that only China's planned cities can do. Qingdao's rapid rise into a prosperous, bustling modern city is why China is so tantalizing for seafood exporters. Canadian, Australian, American, and increasingly Russian planes are landing here with growing frequency, unloading tonnes of chilled shellfish every day.

But politics sometimes gets in the way of what was otherwise a phenomenal success story. U.S. exports dropped off dramatically after June 2018, when then-President Donald Trump hammered China with tariffs as part of his protectionist trade initiatives and Beijing responded with levies on hundreds of U.S. exports, including lobster hauled from the waters off Maine and Massachusetts, adding 25 per cent to their cost. U.S. lobster exports plummeted, especially in Maine, where live lobster exports to China collapsed by 81 per cent between June 2018 and the same month a year later.

The 2018 China-U.S. trade war created a boom for the Canadian lobster industry, driving up the price fishermen were getting by $2.50 a pound in twelve months. By the fall of 2019, with Chinese New Year celebrations looming, lobster exporters across Canada's Maritime provinces were working around the clock to meet new orders from China. Exports to China more than doubled in the year after the U.S. trade war began, which meant companies like Nova Scotia's Tangier Lobster had to hire extra staff to keep the trucks running on time. When I visited that fall for a story for *The Globe and Mail*, workers in white coveralls were working rapidly to pack three thousand pounds of live lobsters into insulated boxes from a nondescript warehouse tucked down a dirt road about an hour and a half outside Halifax.

By the end of 2024, China was importing 24,480 tonnes of live lobster from Canada—a three-fold increase from the volumes in 2016—making it the largest market for live lobster in the world, according to data from the Lobster Council of Canada.

"Increasingly, our focus has been China, China, China," Stewart Lamont, managing director of Tangier, told me. Among lobster exporters in Atlantic Canada, he's considered something of an elder statesman, and his company buys from a network of more than forty lobster dealers around the region. "China is a market entrant like no other, and we've never seen anything like this before."

China, with its seemingly endless appetite for the world's seafood, has often used lobster as a weapon in trade wars. Australia was caught in a prolonged trade battle with China, a conflict that expanded in 2020 after then–Prime Minister Scott Morrison publicly supported an investigation into the origins of COVID-19. Furious Chinese officials responded with a series of tariffs, bans, and restrictions on coal, barley, copper, sugar, timber, wine, and lobster. Australia's $680-million U.S. rock lobster trade with China immediately ground to a halt as Chinese customs officials began quarantining incoming shipments.

The prospect of a trade spat often sparks anxiety in the Canadian lobster sector, which watched how the country's soybean exports to China essentially stopped in late 2018 after Canada's arrest of Huawei executive Meng Wanzhou on a U.S. extradition warrant. Canada's largest soybean market effectively vanished as Chinese officials subjected Canadian soybeans to lengthy testing at Chinese ports in what was widely seen as a retaliatory measure. Chinese buyers started looking elsewhere for product, and Canadian soybean exports to the country declined by 98 per cent.

But in 2023, at least, few people at the seafood trade show seemed worried about the prospect of another trade war. There were deals to be signed and money to be made. Set all this to a backdrop of declining catches of lobster in Canada's most southerly waters, the U.S., and in Europe, and you can guess where it's all going. Lobster as a poor man's food, as it was regarded just a few generations ago in North America, seems like a quaint memory.

"The price just goes up," Alex says, shrugging. "That's what it always does."

17

The Yellow Sea

NANTUN VILLAGE, China – Fishermen stained brown by the sun and with cigarettes clenched tightly between their lips wade into the water at low tide, carrying nets on their shoulders. Wooden long-liner boats, protected from the sea by a narrow stone pier, bob in the gentle waves. Women with their heads wrapped in scarves wash tiny squid in tidal pools on the beach and sell them to passersby from plastic bins.

The only thing that's changed in this scene in the past century is the addition of the slow, methodical chugga-chugga-chugga of diesel engines pushing the boats through the water. Otherwise, the work at the Nantun Wharf continues much as it has for centuries. Some of the fish caught here will be dried, to be sold at the outdoor market a short walk away, put on offer among the crowded vendors' tables piled high with live chickens in cages, nuts, dried fruit, and spices.

The Yellow Sea, the silty body of water that separates the Korean Peninsula from mainland China, has long been a source of food for the people who live around it. The men who keep their boats at Nantun Wharf are after what they've always sought—red sea bream, fleshy prawns, and small crabs, caught in long, rectangular traps. But their catches have been declining for years. Combined with concerns about

contamination and overfishing, it's left consumers along this coast looking to the other side of the planet to source their seafood.

For thousands of years, the Yellow Sea provided food for the millions
of Chinese and Korean people who live near it. But industrial pollu-
tion, agricultural runoff, and domestic sewage have contaminated the
sea's shallow coastal waters, which used to be full of fish life, including
Pacific herring, Japanese mackerel, and cod. It's estimated that about
40 per cent of the Yellow Sea's tidal flats have been developed by humans
in the past century, habitat loss that, combined with overfishing, has
pushed this small corner of the ocean to the brink.

If the scene at the Nantun Wharf is representative of Old China,
when nearly all food was local, then New China is just down the beach.
Young women pose for selfies on the stairs leading away from the pier.
A surf-themed coffee shop sells lattes for 42 yuan, or about $8 Canadian,
along with pizza and gourmet burgers—something the young barista's
parents likely never ate at her age. All around Golden Beach are signs
of China's new wealth: families crowded into luxury hotels that dot
the waterfront, teenagers lining up for speedboat rides, and gleaming
Mercedes parked along the boardwalk.

This China wants the good life, partly fuelled by a generation of
international students who have gone abroad and returned home with
more worldly, and expensive, tastes. And few foods represent China's
arrival at the global luxury food table quite like lobster.

"Before, it was a rich man's food. Now everyone in China can enjoy
the good life," explains Ziyi Ye, a twenty-something logistics manager
for a seafood company in Qingdao.

When it comes to imported seafood, crustaceans like shrimp, crab,
and lobster are the preferred species among Chinese consumers, partly
because their own fisheries can't harvest enough. In 2022, China
imported $5.65 billion U.S. worth of warm-water shrimp, most of it
farmed and coming from India and South America, at a time when sales

into other key shrimp markets, including the U.S., Europe, and Japan, all declined.

But while shrimp and crab may be popular in China, lobster is put on a pedestal like no other. Not far from Golden Beach, the Hilton hotel rises on a hill overlooking the sea with European elegance. Its soaring ceilings look inspired by Buddhist temples, but the ornate chandeliers, jade pillars, and marble fireplaces tell you this is a temple to luxury, not religion. The hotel is a preferred choice for affluent Chinese couples who want to book weddings where lobsters are ordered by the thousands and waiters in black bow ties cater to every desire. The hotel's massive banquet kitchen pumps out endless plates of lobster yee mein, a noodle dish of lobster wok-fried with garlic and green onions and served in a rich brown gravy, a meal typically served at auspicious occasions.

"Lobster is especially popular for weddings and other celebrations because it's an auspicious colour," explains Ji Peng, general manager of Everich Import & Export, a Halifax-based shipper of Canadian lobster. "It turns red when you cook it, and for Chinese people, that's a colour of good luck, of richness, of happiness."

If you want a more down-market dish, you have to walk down the hill to a row of seafood restaurants across from the beach that proudly put images of lobster on their signs. After walking around the trade show for several hours, I am ready to eat. I point to the lobster and nod.

Ding Xiaonao, the restaurant's owner, a smiling woman with a short haircut and a fuzzy black sweater-vest, urges me to come sit inside. Using my phone to translate into Mandarin, I manage to tell her why I am there and what I want. She surprises me by speaking a few words of broken English. "Yes, yes, we have lobster," she says. "It's *Boston* lobster." Ding's accent turns the word into "Bush-ton," but that doesn't matter. It's actually a Canadian lobster, sold by a local seafood distributor. And it's by far the most expensive meal on her menu, about three times the price of the crab.

Ding turns the preparation of the lobster into a spectacle worthy of the price. She begins by taking an orange net and scooping a feisty creature from a wall of fish tanks bubbling away with live crabs, aggressive mantis shrimp, oysters, clams, eels, and slippery octopus. Satisfied the lobster is alive and healthy, she rinses it with a nearby hose, and her sons carry it over to the table in a stainless steel bowl. After making a big show of cutting the rubber bands that hold its claws, they immediately steam it in a propane-fuelled cooker built right into the middle of my table. When it's done, Ding uses scissors to cut the tail meat free from the shell, then dunks it in a bowl of tangy soy sauce with green onions and cilantro. She disappears for a minute to crack the claws, and returns with two more dipping sauces—one a spicy mixture packed with red chilies, the other a fiery sauce made from curried ginger and hot peppers. I ask her if she sells a lot of lobster this way.

"It's not bad," she says with a shrug.

North American lobster companies that have seized on China's lobster craze have enjoyed significant growth because of it. Until 2017, Bayshore Lobster and Seafood, a packing plant founded in St. George, New Brunswick, in 1994, had very little business in China, sending most of its lobster to the U.S. and Europe. But then new owner Nathan Song began aggressively courting business there by investing in live shipments. In 2023, the company shipped three and a half million pounds of live lobster to China. It expects to increase that to four million pounds soon, says Elley Chen, the company's marketing manager, although she acknowledges the Chinese market is getting more competitive for exporters throughout North America.

"Most of our growth has depended on the Chinese market," Elley tells me. "The Chinese want quality, and they want hard-shelled lobster. The restaurants know what they're looking for, and they'll take bigger lobster that they won't in Europe. But it's become a really competitive industry. Everyone wants to sell there."

Canadian lobster companies have an advantage, she says, because they primarily have a winter fishery that harvests when the crustaceans have their hardest shells and the most meat. Their American counterparts often fish in the summer, Elley says, when lobster are shedding, and those soft-shelled versions aren't well suited for the long journey to Asia and spending weeks on display in someone's restaurant. But the Americans have the upper hand with live lobster when it comes to logistics, she says, because they have more freight companies on the East Coast that can get lobsters to China within the sixty-hour window that's required to keep mortality down. Any longer than that, sitting on the tarmac or packed in a warehouse waiting for space on a plane, and the shellfish begin to die.

It's critical that lobsters arrive in China healthy and active, because Chinese buyers often want to see their seafood swimming around, as proof of its freshness, before they eat it.

"People like to select the one they want, and have the chef cook it for them," Elley tells me. "They feel that's the best, and the freshest, way to eat lobster."

Indeed, many of the small seafood restaurants that line the streets of Qingdao's old downtown display their menus in the form of live tanks just off the sidewalk. Diners can pull up a plastic chair, order a glass of beer, and point to the entree they want as it obliviously squirms and swims in the fish tank.

Europe, which was the first place to start buying planeloads of North American lobster in the 1970s and '80s, is a more matured market with less room for growth, Elley says. Buyers there are more selective, wanting only smaller lobster. But Chinese buyers want all sizes, especially big ones, she says, and they buy year-round—not just at Christmastime and the summer holiday season, like importers in France, Italy, and Spain. As well, Chinese consumers are used to buying everything online, and e-commerce companies such as Alibaba have seized on the lobster craze by offering delivery directly to people's doors, something rarely seen in Europe.

For an economy that has, before slowing down after the pandemic, grown at a head-spinning rate, often with GDP growth in double digits—far outstripping anything seen in the West—lobster is just another symbol of prosperity.

"Chinese consumers can afford luxury now, and lobster tastes like luxury," Elley says.

China's global campaign to import more of the shellfish has fuelled allegations in some parts of the North American industry that China has too much control over the buying and selling of lobster. Under both U.S. and Canadian law, only independent owner-operator fishermen are allowed to harvest lobster. But there are no foreign-ownership restrictions on the sale of existing onshore plants, and owners with Chinese roots have been buying up those businesses rapidly. Complaints about growing Chinese influence on the lobster supply can be heard on wharfs and in some corners of government.

Rick Perkins, a Conservative member of Parliament from Nova Scotia's South Shore, was a member of the parliamentary committee looking into foreign ownership and corporate concentration of commercial fishing in Canada in 2023. "My concern overall," he said in one meeting, "is the growing influence of China and the control of our lobster industry itself and that's throughout the supply chain. They're doing it through the back door what they couldn't do through the front door, which was basically own the actual fishing licences. They can't do that, so they're trying to control the buying and the export at the airport."

But companies with Chinese origins bristle at the suggestion they're somehow undermining the Canadian lobster industry. First Catch, a Chinese venture that came to Halifax in 2016 and has become one of the largest airfreight exporters of lobster in the province in just a handful of years, says Rick's suggestion that it prioritizes shipping for Chinese-connected seafood buyers is untrue. The company spent more than $9 million Canadian on its cold storage facility at the Halifax airport and signed a twenty-five-year lease at a new logistics park where live lobsters are packed for the long flight to Asia. It can hold close to ninety thousand pounds of live lobster for extended periods of time, the company says.

The company developed innovative ways to keep lobster alive longer while they wait for shipment to its facility in Changsha, a city of ten million in central China's Hunan Province. That includes a so-called lobster shower, which keeps the animals cooled under a cascade of water from a 378,000-litre reservoir that was built under the facility.

Investments like these are good for an industry that sold very little lobster into China just fifteen years ago, says Geoff Irvine of the Lobster Council of Canada. And as long as Chinese people want to buy North American lobster, there will be people willing to sell it to them.

"Young, wealthy Chinese people want to show their wealth by buying a big lobster and sharing it. It's a status symbol," he says. "So that's great for us. We have premium product that seems made for that. This is market we built up thirteen years ago from a million dollars to more than $400 million."

But as important as the Chinese market has become, the lobster industry is constantly looking for new markets, trying to diversify away from its dependence on the behemoth that is China, as protection from the constant threat of more tariffs and trade wars in the future. The Lobster Council is working to develop new customers in other Asian countries such as South Korea, Singapore, Taiwan, Vietnam, Thailand, and the Philippines, where a growing affluent middle class is increasingly looking for luxury foods from the other side of the planet.

The list of international seafood campaigns goes on and on, from Dubai and across the Middle East to eastern Europe and beyond.

Meanwhile, Geoff says, back in North America a growing taste for meat-less diets means lobster's future may increasingly be in foreign markets.

"Developing new markets can be a challenge, but not as big a challenge as we have facing us in North America. It's hard to predict what Gen Z is going to demand, with this beyond-meat movement," he says. "Are we going to have genetically modified lobster someday? And what's that going to look like? These are the kind of questions we're asking."

18

The Music of the Sea

NEW QUAY, County Clare, Ireland – Gerry Sweeney leaves the diesel engine rumbling on the little green-and-orange boat and starts unloading the plastic tubs of velvet crabs onto the wharf against a slate-grey sky. As he carts his catch toward his fish shop at the end of the stone pier, one crab falls out of the bin and crawls to freedom. The fisherman glances at the crustacean scurrying away and smiles.

"It's his lucky day," he says, and keeps walking.

If you're an Irish lobsterman, there's no time to waste. Gerry needs to pack a load of lobster and crab bound for Spain, a three-day trip in a vivier truck that will keep the shellfish alive in seawater for the journey. It leaves tomorrow morning.

Gerry, sixty, has been fishing the sheltered waters of Galway Bay in western Ireland for forty years. He usually fishes by himself, because there's not enough money in lobstering anymore to hire a deckhand. His sons, who used to help him in the summers, have little interest in continuing in this line of work.

His boat the *Ceol na Mara*, which means "Music of the Sea" in Irish Gaelic, looks like a time capsule from another era as it sputters away at the wharf, a fraction of the size of those modern multimillion-dollar boats that fish off New England and Atlantic Canada, but there's not enough revenue coming in to invest in the vessel. Used tires are strapped to the boat's side to protect it from rubbing against the wharf, and the thirty-year-old *Ceol na Mara* could use a coat of paint. Gerry doubts anyone will want to buy it, or continue to haul traps in these waters, once he's gone, anyway.

"They're all lined up now," he jokes, gesturing toward the fish store's empty doorway.

On Ireland's rocky, rugged western coast, the lobster fleet has dwindled to just a handful of fishermen. They all know each other by first name, and have carved out their own solitary turf along the shoreline dotted with the ruins of medieval castles and stone wharfs that few other people use. A laneway lined by rock walls and barely wide enough for two cars to pass leads down to the pier at New Quay, where Gerry keeps his boat. The water here is sheltered by a peninsula lined with cattle pastures and flanked by treeless hills. It's good conditions for oyster farming, something people have done for centuries.

It also used to be good lobster fishing on these protected waters, and fishermen so inclined could haul their traps year-round. But Gerry, sorting crabs inside his shop's live storage room decked in yellow overalls and purple gloves, says overfishing has drained the bay of its once plentiful lobster population. He now puts seven hundred pots in the water—Ireland, unlike Canada and the U.S., has no limits on the number of traps fishermen can use—but catches less than he did decades ago when he used only ninety traps.

"There's no good years anymore," he says, although he declines to say just how many lobsters he's catching now.

In the 1870s, two decades after the Great Famine, more than five thousand boats were fishing for lobster in Irish waters, and the industry employed some twenty-three thousand men seasonally. In that era, lobster was so plentiful it was fed to pigs and goats and used as fish bait. It was traditionally a poor man's food in a nation that always preferred lamb and beef. (Irish fishermen catch a spotted dark-blue-shelled variety, *Homarus gammarus*, known as European lobster, similar in taste to its North American cousin but about twice the price.)

The island's lobster harvest peaked in 2004, when fishermen reported annual landings of nearly 900 tonnes. Today, it's a shadow of that amount. Ireland's lobstermen reported catching a mere 146 tonnes in 2021, according to Bord Iascaigh Mhara, the country's fishery development agency.

For fishermen who have worked Ireland's waters for decades, the rapid decline in the catch in the past decade has been staggering. The declines here mirror the general drop in catches happening in all lobster fishing nations in Europe.

"I don't know if there will be a fishery left after me," Gerry says.

It's all made local lobster an increasingly rare, and expensive, sight on Irish dinner plates. There's a lobster bar next door to Gerry's fish shop, complete with a logo of a red lobster clutching a pint of Guinness, but few lobster caught in Irish waters are actually eaten by locals. It's too expensive for most. Linnane's Lobster Bar will sell you a pound-and-a-quarter lobster for €42, or about $60 Canadian. Down the road at Monks, a seafood restaurant by the nineteenth-century stone pier in the coastal village of Ballyvaughan, a lobster dinner will cost you over €100.

For a nation surrounded by water, the Irish are strangely suspicious of protein that doesn't come from a meadow. Only about 5 per cent of the lobster caught by Ireland's fishermen are consumed domestically. Part of that is the high price. But there is also the lingering legacy of Fish Fridays, the old Catholic edict that households weren't allowed to eat meat on Friday, which turned many consumers off seafood back in the era of poor quality control and insufficient refrigeration. Today, while the French, Spanish, Belgians, and Italians will pay premium prices for lobster pulled from Irish waters, many Irish wrinkle their nose at the thought.

Though the rules around Fish Fridays were loosened in the 1960s, its influence on food choices and consumer behaviour is still felt around the world today. Fish prices in many countries with large Catholic populations plummeted after the ban on meat was relaxed. And the tradition spawned innovations in seafood, including the invention of the Filet-O-Fish sandwich by Lou Groen, who owned a McDonald's franchise in a largely Catholic neighbourhood of Cincinnati, Ohio, which was struggling to sell burgers on Fridays. McDonald's founder and CEO Ray Kroc apparently hated the idea—he was worried his restaurants would smell like fish—but even he had to eventually admit it was a good seller.

But as a valuable luxury export product, lobster remains one of Ireland's most important shellfish, bringing critical revenue to rural coastal communities that are far removed from the economic bustle of Dublin. As a part of the European Union, the republic also has a leg up on British fishermen across the Irish Sea when it comes to selling lobster to buyers in continental Europe.

But even as the fishery declines, the Irish government doesn't actually know how many lobsters are being caught in its waters each year. Many argue the stock is in worse shape than it appears, but overfishing is hard to track. That's because, under European Union rules, boats under thirty feet don't need to report their catches. Much of the fishery is still done on these smaller vessels, meaning those landings aren't declared at all.

"That's the kernel of the problem. We don't have an accurate picture of what's actually happening," says Séamus Breathnach, a lobster fisherman in Carna who used to work for Bord Iascaigh Mhara.

That's a big issue for an Irish fishing industry that sets no limits on the catch and doesn't limit access to the fishery. There are no controls to regulate overfishing or any mechanism to adjust for the increased fishing effort that's bringing back dwindling returns to the dock.

"We're working harder and harder just to stay in the same place," Séamus says. "We've tried to get a proper management system for years, but the regulator isn't interested. And industry hears 'management plan' and all they think is catch limits."

In Ireland, the V-notching system intended to protect egg-bearing females is voluntary, and fishermen are paid only 50 per cent of the market price for lobsters they mark and throw back. For some, there's little incentive to cooperate. Enforcement, while more common for the bigger offshore fleet, is rare in the inshore fishery.

"In the three years I've been back fishing, I've never seen an enforcement officer on the wharf," Séamus says.

Séamus fishes in Connemara, a rugged, rocky, rainy region at the very western edge of Ireland where historically there was little else but subsistence farming and fishing. Like many people in this remote corner of the island, he still speaks Gaelic to his fellow fishermen, who use the same wharfs their great-grandfathers used in the 1800s. In their local dialect, lobster is known as gliomaigh. If you were to set into the sea here and head west, you wouldn't hit land until you reached North America.

His grandfather was a lobster fisherman who used to fish in black-hulled, maroon-sailed boats called Galway hookers, the once ubiquitous workhorse of this area that was preferred up until the 1950s. Some of the earliest lobster fishing around Boston and Cape Cod Bay was done in similar boats built by Irish immigrants recalling the small, utilitarian vessels they had used back home. They called their boats Boston hookers.

They're part of the legacy of Irish fishermen who spawned many multigenerational fishing families that settled along the New England and Atlantic Canada coastlines. Many arrived in North America poor, uneducated, and discriminated against, and fishing was one of the labour-intensive jobs that allowed them to eke out a living. Between 1845 and 1855, when the island's potato crop failed and brought a devastating period of starvation and disease, some two million Irish came to ports in America and Canada. Today, about one in five residents of Massachusetts, New Hampshire, Rhode Island, Maine, and Newfoundland claim Irish ancestry. Not surprisingly, their bloodlines run throughout the North American lobster fishery, along with descendants of English, Scottish, and French settlers, as well as Indigenous fishermen.

Séamus's home wharf in Carna, the Ard West pier, is like a museum for Irish lobster fishing history. To reach it, he carefully guides his little car down a winding single-lane road, shooing away a white Connemara pony that stubbornly blocks his path. He's a "Carna man," as he likes to say, and he proudly points out his boyhood house and his relatives' homes that dot the windswept, treeless landscape where nothing else seems to move.

It is both staggeringly beautiful and incredibly lonely, as if everyone else left long ago. Carna's population before the Great Famine was close to eight thousand. Today, there are fewer than two hundred people in the village and perhaps another fifteen hundred in the surrounding area. The ruins of an Irish chieftain's castle keep watch over the pier and its narrow channel that leads to the wide-open Atlantic Ocean. Because this coastline is so exposed, fishermen here might only get a hundred good fishing days a year.

The progression of the fishery is in plain view from Séamus's wharf. Currachs, the ancient timber-and-hide boats that were once the vessel of choice when lobstering was exclusively a small-scale, hand-powered fishery using wicker pots, sit rotting a few feet away from the edge of the high-tide mark. At low tide, the larger timber half-deckers that replaced the currach rest at odd angles on the ocean floor. Newer fibreglass vessels that can carry up to 150 wire lobster traps are moored nearby.

The Irish may have adopted modern fishing boats and equipment, but they hold on to old traditions that predate anything seen among North America's lobstermen. Every July 16, following the ways of his ancient Celt ancestors, Séamus and his fellow fishermen visit a shrine on nearby St. MacDara's Island to have their boats blessed and pay their respects to St. Sinach MacDara, a patron saint of local seafarers. One tragic pilgrimage in 1907 resulted in the drowning of nine people, as a horrific storm swept in off the Atlantic and devastated the local fleet.

It's believed the saint built a wooden church on the island in the sixth century, and fishermen who sailed past were encouraged to dip their sails in the water three times in reverence. Irish folklore holds that St. MacDara could summon up storms against anyone who didn't acknowledge him. Irish historian Roderic O'Flaherty reported that in 1672 a naval captain from the garrison in Galway neglected to dip his sails, ignoring the local tradition, and was subsequently killed in a shipwreck.

In a part of the world where vicious Atlantic storms regularly send towering waves crashing into the cliffs, fishermen here have long learned to respect the sea and its power. Anything that might offer a little extra protection is worth doing, they reason.

"I'm not a religious person, but I still follow the old traditions," Séamus explains.

There was a time when fishing was the only thing that provided food and money for men in the Connemara region, where the rocky soil is poor for farming apart from grazing sheep and there's little factory work to speak of. The fishermen knew what shoals to avoid and where the channels were most treacherous. Yet with declining returns and long days at sea, few young Irishmen want to do this work anymore, which worries Séamus.

"You're losing part of the heritage," he says.

Another threat to Ireland's fishery is cheaper imports from North America. In 2020, the U.S. negotiated a trade deal with the European Union to eliminate tariffs on imports of live and frozen American lobster for five years, in exchange for the U.S. cutting back tariffs it had imposed on frozen meals and crystal glassware. Ireland's National Inshore Fishermen's Association condemned the deal, saying it would depress prices for Irish-caught lobster, one of the last surviving fisheries for the hard-pressed inshore fleet.

Patsy Mullins, who has been fishing lobster in Galway Bay for more than forty years, warned that a drop in prices would pressure fishermen to put out more traps, further overfishing a species that is already showing serious signs of trouble. The fishery, which could once fetch €32 per kilogram—or about $20 Canadian a pound—for fresh lobster in France during the Christmas peak, was still recovering from the collapse in prices by two-thirds during the pandemic.

"This EU-US deal is really going to hit younger shellfish skippers badly," Patsy told *The Irish Independent* newspaper. "Decent prices allow boats to fish less, and is better for stocks. This deal flies in the face of our sustainable aims." He added, "Lobster fishing has kept money coming into peripheral areas where there are few employment alternatives."

Back at the wharf in New Quay, Gerry Sweeney has no time to ponder international trade deals or worry about cheaper North American imports. He's got a load of shellfish that needs to be packed for shipping. Tomorrow, he'll fill the *Ceol na Mara*'s tanks with diesel and head back out on the bay to check his traps, as he has since he was a young man. Fishing is all he's ever known.

He also knows this way of life won't last forever, and he worries that, without limits on the fishery, the resource could disappear. But his wife, Marilyn, who runs the fish shop with its bubbling live tanks while he's at sea, says he'd sooner collapse than stop plying the waters of Galway Bay for lobster.

"As long as those two legs work, he'll keep fishing," she says. "You can't stop him."

19

The Lobster Capital of Europe

BRIDLINGTON, Yorkshire, England – The old harbour at Bridlington is rimmed with shops selling fish and chips, served up hot and crispy and on the cheap. It should be no surprise that a blue-collar town such as this loves that ever-reliable staple of the English diet, long the closest thing to Britain's national dish. But here, in the self-proclaimed Lobster Capital of Europe, a working-class city of about thirty-five thousand people on England's Yorkshire coast, you'd be hard-pressed to find anyone serving lobster—even though the boats coming into the harbour are loaded with them.

Crates of speckled blue-shelled *Homarus gammarus*, that European cousin of American lobster, are being stacked into the refrigerated trucks that rumble away from the wharf every night, bound for European dinner plates. In their wake, they leave behind the lingering smell of diesel that mixes with the scent of old bait that attracts a small army of well-fed seagulls. Bridlington's port bustles with crab and lobster boats that are moored five abreast when they're waiting for the tide to come in.

They used to catch lobster in wooden boats launched from the long, flat beaches that surround the town, but in recent decades, a sizable commercial fishery has developed around the prized shellfish. Yet

Bridlington, which at three hundred tonnes a year lands more lobster than any other port in Europe, is hardly a mecca for people who actually want to eat the shellfish.

If you walk into the nearby Tesco, the British-owned grocery chain that is one of the largest multinational food companies in the world, you'll find lobster—frozen, cooked, and from Canada, stacked in boxes in the freezer section right next to the turkey and the fish sticks. A four-hundred-gram lobster will set you back £11, or about $17 Canadian. There are no live lobster tanks to be found, and the frozen variety aren't exactly flying off the shelves.

"I've never seen anyone buy one," one of the clerks told me with a shrug.

At the Lobster Pot restaurant, a popular eatery in town, there's no lobster on the menu, either. The manager says it's too difficult to get, and too expensive for their customers. The boats come in to Bridlington's long pier all day, bringing fresh loads of lobster and crab. Yet nearly all of it is immediately loaded onto the big refrigerator trucks that are running around the clock, bound for France.

In Europe's "lobster capital," this is what happens when the price of local lobster rises beyond reach of most consumers while driving fishermen to catch as much as they can. Lobster has become a rare delicacy, shipped abroad to affluent diners, and eaten domestically only on special occasions by tourists and the few British foodies willing to pay handsomely for it. Once a common part of English and Continental coastal diets, European lobster now accounts for only about 3 per cent of all the lobster caught in the world. And the exclusivity of its market serves as a warning for coastal communities in Canada and the U.S. watching catches of North American lobster decline while the price to consumers climbs.

The lobster fishery on England's scenic east coast has tried to cultivate a domestic market for their lobster, especially as Brexit has thrown a wrench into the business of shipping lobster across the English

Channel, adding significant delays and shipping costs. They've had limited success. Taking a page from Maine's successful lobster marketing campaigns, the local fishery is trying to promote "Bridlington Bay lobster," but so far it's mainly tourists who are showing any interest. While there's growing support in larger British cities for "native" lobster, and the pandemic helped encourage Britons to try cooking seafood at home, the high price remains a hurdle. You can pick up a frozen fully cooked Canadian lobster at a supermarket for as little as £9.99, but a cooked European lobster caught just offshore will set you back £61.99—over $100 Canadian. It's not a completely fair comparison, but the premium price for local lobster makes it a special-occasion-only splurge.

To persuade consumers to pay higher prices for lobster from British waters, seafood companies here need to sell their shellfish as a more refined alternative to the North American imports that flood their market. It's necessary for them to dismiss the "cheap Canadian lobsters" that are being stocked in British grocery stores, especially in the run-up to Christmas, as an inferior product.

"We don't do surf-and-turf here. It's much more elevated, more premium, more higher end," explains Julie Hill, sales and marketing officer for the Lobster Pot, a multigenerational seafood exporter based in Church Bay, nestled among the cliffs and pebble beaches of the northwest Welsh coast. "We don't put this in a lobster roll or in a brunch buffet. It'll be in a Michelin-starred restaurant."

But Bridlington Bay lobster remains a hard sell. This is, after all, Yorkshire, where on Sundays people would still rather crowd into pubs or around the family table to eat a traditional roast beef dinner, complete with pools of gravy and crispy Yorkshire pudding. Although archeological digs have found piles of shells, suggesting lobster was once a common part of the diet in English coastal villages, today it's rarely on the menu.

There remains a disconnect among consumers in Britain around lobster, with little political pressure to push for a more sustainable

A man rows lobster traps to shore in Blue Rocks, Nova Scotia, in this 1947 photograph. This old-fashioned method of fishing was being replaced in the post-war era, when the Canadian government gave subsidies to fishermen to build new, more modern vessels.

Dougal Doucette holds up the first large lobster of the season in Miminegash, Prince Edward Island, 1948. The average sale price for lobster that year was 25 cents a pound.

The *Small Fortune's* sits tied up at the wharf in Dipper Harbour, New Brunswick. Larger, modern fishing vessels allow fishermen to go further out to sea and in worse weather. They're increasingly automated, too: "Today, a kid with a Nintendo could do this," the boat's captain says.

The *Small Fortune's* captain, Brad Small (seated), and first mate Tom Duke stuff bait bags with diced fish at the wharf in Dipper Harbour, New Brunswick. There's no shortage of work to be done to get ready to go back on the water.

Freshly caught lobster are sorted and placed in plastic bins on board the *Small Fortune's* on the Bay of Fundy in November 2023. They could be in China by the weekend.

Stevedores at the Fulton Fish Market dock in New York City unload lobsters in 1943. For many decades, lobster caught in New England was shipped to buyers in New York by sailboats called smacks, which had water-filled wells for keeping their catch alive.

Jimmie Liebl, a salesman at the Fulton Fish Market in New York, shouts to a bookkeeper after weighing lobsters in 1939.

The *Sea Bug*'s first mate Andrew Robinson watches as the boat is guided into the wharf at Saulnierville, Nova Scotia, in September 2020. A police barricade prevented non-Indigenous fishermen from entering the wharf as tensions grew over the unregulated Mi'kmaq fishery.

The Mi'kmaq crew of the *Sea Bug* measure lobster and attach rubber bands to their claws on St. Marys Bay, Nova Scotia, in September 2020, the same month protests exploded over the Indigenous fishery.

Crowds pile out of the China Fisheries and Seafood Expo in 2023. China's appetite for the world's seafood seems to know no bounds. Between 2018 and 2022, the country imported between sixty-two million and seventy-two million pounds of lobster annually.

Small fishing boats are tied up at Nantun Wharf, on the Yellow Sea in China. Overfishing and polluted waters have led consumers in China to look further abroad for their seafood.

Gerry Sweeney's boat, the *Ceol na Mara*, sits at the pier in New Quay, County Clare. The lobster fleet on Ireland's rocky western coast has dwindled to just a handful of fishermen. "There's no good years anymore," he says.

Séamus Breathnach, a lobsterman in Carna, Ireland, is worried his country is losing its lobster-fishing heritage as catches decline and fewer young people get into the industry.

A man inspects confiscated lobster pots on Valentia Island, Ireland, at the turn of the century. Poaching remains an issue in the modern lobster fishery today.

Lobster traps sit on the wharf in Bridlington, U.K. While this port catches more lobster than anywhere else in Europe, few locals eat the shellfish because of its high price. Most of the harvest is exported to restaurants in France, Spain, and Italy.

Lobster and crab boats are tied up on the waterfront in Scarborough, U.K. Despite signs of overfishing along this coast, many fishermen here bristle at the suggestion of any restrictions on the number of traps they can use.

Library and Archives Canada

Women pack lobster meat inside a cannery on the Miramichi River, New Brunswick, in 1936. Prior to advances in refrigeration, air freight, and modern logistics, most lobster shipped around the world came in a can.

Provincial Archives of New Brunswick

Men scoop lobster out of a floating lobster pound in Bouctouche, New Brunswick, in 1955. As ocean temperatures have warmed, the modern lobster industry has moved to indoor, temperature-controlled storage facilities.

fishery. In a place where live domestic lobster can be two or three times the price of frozen lobster imported from North America, the local option is seen as a fancy shellfish destined to leave the country. It's for the French or the Spaniards, not us, the British seem to be saying.

"British people are pretty conservative with their seafood tastes," Bryce Stewart, an affable Australian-born marine ecologist and fisheries biologist who teaches at the leafy brown-brick campus of the University of York, says. "They like white fish. Lobster is very much seen as a luxury, something you only buy at Christmas. For the average person, it's just too expensive. It's a hard sell for people."

There are troubling trends in landings data and other stock assessments that suggest European lobster is becoming an increasingly rare commodity in these waters. Despite efforts to popularize it, Britain's "native" lobster is only likely to become more scarce and expensive.

A 2019 Norwegian study suggests that improvements in fishing technology that make fishermen more efficient at catching lobster are masking a serious decline in lobster populations in the regions where *H. gammarus* used to be most plentiful. The study notes that, even with improvements in fishing equipment, boats, and other gear, there was a 92 per cent decline in European lobster numbers between 1928 and 2019. That's a serious issue for countries like Norway, which has long relied on the coastal lobster fishery. The Norwegians began exporting live lobster to central Europe as far back as 1650, but didn't introduce any kind of fishing regulations for more than two centuries. Despite modern conservation efforts such as minimum size restrictions and further closures of the fishing season between July and September, the country's lobster fishery has declined to a historically low level.

There's similar evidence of overfishing in the U.K., which is the largest harvester of European lobster. Data shows that while landings in British waters have been in decline for more than a decade, lobster are being found in fewer areas, too. At the same time, the average size of the lobsters being caught is shrinking, another sign of overfishing. But getting an accurate picture of the real impact of the fishery remains difficult.

"There's lots going on, and we don't quite have a handle on it," says Jamie Robertson, managing director of the Bridlington-based Holderness Fishing Industry Group, a nonprofit trade association whose members include fishermen, landing companies, processors, and others directly involved in the shellfish business along England's Holderness coast.

For decades, British fishermen hauled up lobster without an accurate system to measure how many they were taking from the sea. That's a huge problem if you're trying to conserve a species that has been a part of the coastal fishery since before the Middle Ages. And since about 80 per cent of all the European lobster caught every year come from the British Isles, what the fishery does here has impacts on the entire species.

Jamie, a bearded former commercial fisherman who's built like a rugby player, remembers the old artisanal lobster fishery of his grandfather's era. Fishermen at that time used wooden cobles, traditional flat-bottomed, single-mast boats that they would haul out of the water by horse and later tractor. They hand-wove wicker baskets for lobster pots, and hauled up their catches by hand, never dreaming of today's hydraulics that have made the fishery more efficient. The traditional Yorkshire coble is preserved by historical societies today, although a few diesel-powered versions are still used by local fishermen. Most lobster fishermen here have switched to modern vessels that are going farther out into deeper sea, reaching previously unexploited lobster grounds, and use more pots.

Ports like Bridlington used to be full of fishermen who caught white-fleshed fish like cod and mackerel. But as those species became over-fished, they seized on lobster as the next boom—and began catching them at unprecedented levels. Until the late 1990s, lobster fishing here looked very different from how it does today. Individual hand-hauled traps were replaced by long strings of traps and motorized haulers, which exponentially expanded the amount of lobsters fishermen could catch every time they went out. The number of traps grew in response, boats got bigger, and their crews ventured farther from shore.

"It's gone from a small scale, artisanal, to a full commercial scale, and all for export," Jamie says. "The continental market has driven that growth, because fishermen can get more for their lobster in Europe. Traditionally, lobster was always caught by hand here, one at a time, and that slowed things down. So everything has changed."

Across the English Channel, the French fleet has nearly fished European lobster into oblivion, with only a handful of fishermen left who are still catching lobster on a small scale along the country's northwest coast. The French have long loved their "Bretagne" lobster, but today, most of the lobster available in France under that name has almost certainly been imported from England, Ireland, or Scotland and sold throughout the country by the old established shellfish dealers who still operate there.

Europe's famous blue-shelled lobster, typically sold for about twice the price of its North American cousin, is sure to become even more of a premium product, available to only the most affluent of diners, as it becomes more scarce. That's a problem for the cafés and high-end restaurants in Paris, Rome, Barcelona, and elsewhere, where the European lobster has long been a symbol of luxury. And yet limits on its harvest are almost nonexistent in England—while Canada allows a maximum of 350 lobster traps per lobster licence in the majority of its fishing zones, and most jurisdictions in the U.S. limit fishermen to 800 traps, few fishing zones along the English coast place any limits on the number of traps a fisherman can use. Some fishing crews are putting as many as eight thousand pots in the water. A rotation of pots that large would take six to eight weeks to haul up, meaning some lobsters could spend over a month stuck in traps, leading to deaths and lost limbs as the trapped shellfish turn on each other.

And that kind of damage to the fishery isn't recorded by landing statistics. With no processing industry for lobster in the country, dead or one-clawed lobsters are worthless to a fishery that wants only live whole lobsters, so they're tossed overboard, and are consequently missing from official records. Industry watchers know it's happening, but can only guess at the numbers.

"We don't even know how many pots are in the water," says Michael Roach, a fisheries scientist for the Holderness Fishing Industry Group, which maintains a research vessel and an office at the end of the long pier in Bridlington. A fish tank bubbles away in his second-floor office as he talks.

There's little diversity in the fishery, or even economic incentive to catch anything else, Michael says. Legislative and regulatory barriers make it difficult for fishermen to rotate between species, so they just try to catch as much lobster as they can. Most white fish still being caught in British waters are harvested by company-owned vessels, leaving lobster and crab as one of the few profitable independent fisheries that a fisherman can enter on his own and work for himself.

But even as lobster stocks decline in British waters with calls for urgent action by both the Marine Conservation Society and the U.K. government, there's not a lot of support for further restrictions on the catch. Chris Townsend runs a crab and lobster wholesale business from the long Victorian-era brown-brick warehouse on the Scarborough waterfront, just up the coast from Bridlington. Forklifts buzz up and down the pier, loading the local catch into refrigerator trucks idling nearby. Unlike most wholesalers here, Chris sells all his lobster in the U.K., to restaurants in the bigger cities of London, Manchester, and Sheffield.

He bristles at the suggestion that the local lobster population is being overfished. He says most British lobstermen are doing a good job of self-regulating, and any talk of limiting the number of traps would only cut into their livelihood. He says few fishermen are using eight thousand traps; most are closer to twelve hundred.

"That's like cutting your income in half," Chris says, working in rubber boots and a T-shirt on an unusually warm October afternoon. "And the bigger question is, how would it be policed?"

Bryce Stewart says it will take a crisis before British regulators begin to take serious steps to protect the lobster fishery. But he's encouraged that attitudes are slowly changing, as his work shows that there are big changes happening in English waters that are still not fully understood.

"My experience in fisheries management seems to be that things need to go quite wrong before we get serious about making them right," he says. "Whether it's the collapse of cod on the Grand Banks or whatever, it seems to take a bit of a disaster before people get really serious. But things are changing. The recognition for sustainability is infinitely higher than it was ten years ago."

Some conservation measures are slowly being introduced. One fishing zone in the U.K. only recently banned catching egg-bearing females, a practice that has been outlawed in North American waters for decades. Another fishing zone, stretching from Newcastle north to the Scottish border, was the first in the country to put a limit on the number of pots, in this case four hundred. But there remains resistance in the fishery to any effort to scale back the harvest. In Bryce's work, he talks regularly to fishermen from Scotland down to England's southern coast. They share one thing in common: a willingness to overlook their own impact on lobster populations and focus the blame elsewhere, whether that's offshore wind farms, underwater pipeline projects, or the so-called super boats that have large crews and maintain thousands of lobster pots at a time.

"They know that there's a problem, they know they're fishing too hard. But there's a tendency to blame someone else for that," Bryce says. "The average boat is saying, 'Well, I fish a smaller number of pots. It's the *other* guy. These super boats, they're the problem, and that's what we need to sort out.' The small boats are not going to back down. They think if restrictions are put in, they're just going to lose out."

20

The Prince of Brittany

PARIS, France – There are no lobster rolls on the menu at the Restaurant Guy Savoy, where white-gloved waitstaff allow you to feel like royalty for a few hours and the wine list is so big it requires its own special stand. Here, in the epicentre of Parisian fine dining, lobster is elevated to art. Guy's version of the crustacean features delicate moulds of lobster tartare cured in lobster vinaigrette with lobster carpaccio and lobster coral pancake, all seasoned with more lobster vinaigrette. Lobster is the centrepiece of a twelve-course meal that goes for the princely sum of €680—north of $1,000 Canadian—per person. Guy, a Michelin-starred chef who mentored Gordon Ramsay, has made a career of creating lobster dishes so delicate and so pretty that desecrating them with a fork almost feels like a violent act.

From roast lobster with bouillabaisse sauce to lobster with "hot and cold" bouillon, the French chef does not hide his love for this shellfish, nor the fact that he feels his well-heeled guests should always be prepared to pay handsomely for it. It should be no surprise that when Restaurant Guy Savoy moved in 2015, it chose a 4,300-square-foot space in the Monnaie de Paris, the old French mint, with its views of the Seine River.

"And then there are the prices. With these, Savoy makes no attempt to hide either his ambition or his status," the *Financial Times* wrote in 2018.

Since the Middle Ages, the respected place of lobster in French kitchens has not wavered. No country has done more to elevate the status of this shellfish as a luxury food. King Louis XIV, a lover of all things decadent, adored them and frequently had lobster served at banquets at Versailles. In 1571, when King Charles IX staged an elaborate ball to welcome his bride, Elisabeth of Austria, he placed a stack of cooked lobsters on the head table to show off his wealth and power. In 2023, when Britain's King Charles III made his first state visit to France as monarch, blue lobster was once again on the menu. And of course it was a French head servant, François Vatel, who in 1671 famously "died for want of lobster sauce," when he was said to have committed suicide because seafood for an extravagant banquet for two thousand people in honour of Louis XIV wasn't delivered in time. Or so the story goes.

The French aristocracy have continued to hold lobster on a pedestal. It should be no surprise that in Brittany, the home of France's small lobster fleet, chefs are still inventing ways to serve their prized blue shellfish, infusing it with apples, Chartreuse liqueur, kabocha squash, espresso, and smoked shells to create expressive new flavour profiles.

In North America, lobster is typically enjoyed more plainly and with little fanfare, steamed whole and served with butter and perhaps a slice of lemon. While expensive compared with other seafoods, it's still the food of wooden takeout stands, plastic bibs, and sun-bleached picnic tables, often tossed in butter or mayonnaise, wedged into a grilled hotdog bun, and served with fries. Tourists will stand in line for over an hour at lobster shacks throughout New England and Canada's Maritime provinces to get a taste of these pricey but simple meals.

Across the Atlantic, however, French chefs have long taken the shellfish to a higher place. Pierre Fraysse, who had spent several years working in Chicago in the mid-1800s, is credited with inventing a haute cuisine dish he called homard à l'américaine, a now classic recipe featuring chunks of lobster meat served on a plate or piled in the shell, bathed in a sauce that combines tarragon, butter, and tomato, all flambéed with a healthy splash of cognac. The story goes that Parisian chefs disparaged the name, arguing that Americans had no gastronomic culture, and changed it to homard à l'armoricaine, after the region in northwestern France where blue lobsters are caught, and it stuck.

Two centuries earlier, François Pierre de La Varenne, author of the wildly popular cookbook *Le cuisinier françois*, recommended boiling lobster in seasoned bouillon and serving it with vinegar and parsley. He offered a fancier version, too, suggesting lobster pieces could be fricasseed in a white sauce consisting of butter, parsley, verjus (unripe grape juice), egg yolks, and nutmeg.

Upper-class French began eating lobster more frequently in medieval times, thanks to the influence of the Vikings, who introduced more seaworthy boats, which led to an overall rise in the consumption of seafood. Unlike fish, however, unrefrigerated lobster needed to be cooked within two days of leaving the water, so it was most available to those who could afford transportation, according to food writer Elisabeth Townsend. References to lobster and its increasingly important role in European cuisine began appearing in the early 1300s, with instructions for servants to boil it in water and wine and serve it with vinegar.

The most well-known French contribution to lobster cuisine, of course, is probably lobster thermidor, with its rich sauce of heavy cream or béchamel, wine or brandy, mushrooms, stock, Dijon mustard, shallots, and cheese, tossed with the lobster meat and stuffed into a halved lobster shell, topped with Gruyère or Parmesan cheese, and broiled until brown. There are two competing stories about the origins of this succulent dish: One claims it was created for Napoleon Bonaparte, who named it after the month of Thermidor, between July 19 and August 17, when the dish was served to him. More likely, the dish was created by the Paris restaurant Chez Marie and named in honour of a short-lived and controversial play about the French Revolution.

It was an eminent French zoologist, Henri Milne-Edwards, who first classified American lobster as its own species, in his three-volume *Histoire naturelle des crustacés* in 1834. He named it *Homarus americanus*. Henri was a giant among a group of French zoologists who made so many advances in marine biology at the National Museum of Natural History in Paris. He was a pioneer in his field, insisting on the need to study animals alive in their natural habitats, instead of just examining their corpses in museums. So great was his determination to advance marine science that he is mentioned no fewer than five times in Jules Verne's classic science fiction adventure novel *Twenty Thousand Leagues Under the Sea*.

It is sometimes said the French love lobster so much that they were willing to go to war over it. In the early 1960s, the French fleet was getting desperate for new fishing grounds after the loss of much of France's colonial territory in Africa. Parisian diners still demanded their lobster, and the country's fishermen were travelling farther and farther abroad to find it. When they began fishing for spiny lobster a hundred miles off Brazil's northeast coast, Brazil saw it as an act of hostility. Both sides sent warships to escort their fishermen and took their fight to an international tribunal. France argued that lobsters "swim," making them fair game for an international fleet, while the Brazilians said lobsters crawl along the continental shelf, making the shellfish theirs alone. The dispute wasn't resolved until Brazil unilaterally extended its territorial waters to a two-hundred-nautical-mile zone, in order to claim the disputed lobster bed.

"The attitude of France is inadmissible, and our government will not retreat. The lobster will not be caught," Brazilian foreign minister Hermes Lima thundered, according to an account of the dispute in the U.K.'s National Archives.

But the French love for lobster, as ingrained as it is in their culture, cannot do anything to slow the effects of warming ocean waters on their most cherished seafood. Far from the glamour of Parisian restaurants, Cédric Delacour, a bearded fisherman in his early thirties, is one of a small fleet of small-scale lobster fishermen in the northwestern corner of France who are noticing big changes in the water they work on. When the weather is good, he can often be found steaming out from the harbour in Cherbourg in his little navy-blue-and-white boat, *Manola*, emblazoned with an artist's rendering of a naked woman on the bow, with his dog by his side. For centuries, people like him have worked the coves and bays of the French side of the English Channel to catch the local delicacy known as Breton lobster. Here, they affectionately call it the Prince of Brittany, but it's the same blue-shelled European lobster British fishermen are catching on their side of the channel.

But lately there's a troubling trend being talked about by those who catch and cook the prized crustaceans. Cédric told the BBC's Emily Monaco in 2023 he's noticed female lobsters are producing eggs earlier in the season, a sign of stress and warming waters in the channel—which is about 0.6 degrees Celsius warmer than it was a decade ago, according to British marine monitoring labs. The old seasonal rhythm of lobster reproduction that his grandfather taught him has been replaced by unpredictable patterns. It's now common for lobster off the coast of France to produce eggs three times in a year, instead of twice.

"It's a sign that the species feels endangered," he told the BBC. "And so, it lays more."

Warming waters aren't the only sign of a problem. The French fleet's catch of its beloved European lobster has been in decline for more than a decade, after it peaked at 869 tonnes in 2010. Fewer coastal areas are productive lobster grounds these days, and the fishery has been reduced to a small northwestern corner of the country. France, the second-largest harvester of European lobster after the U.K., has had to go farther afield and rely increasingly on foreign vessels to supply its demand for the shellfish. Overfishing of its local waters has left it a net importer of lobster, much of it by plane from Canada and by live-shellfish trucks from Great Britain.

Meanwhile, warming water off the coast has introduced a new challenge for the declining lobster population—the arrival of predatory octopus, appearing in such numbers that fishermen are worried they could decimate the industry. Octopus are another example of unbalance gradually appearing in the ocean's ecosystem as temperatures rise. Much like the arrival of black sea bass in traditional lobstering regions in southern New England, the octopus with their boundless appetites have been destroying lobster catches in the waters around Brittany.

"The octopus got into our trap, and ate the whole catch," fisherman Patrice Douaré, of Quiberon, complained to a local TV station in 2021. "So far, half of my lobsters have been eaten, so of course, at this time of year when the price is rather high due to tourism, it obviously represents a loss of earnings."

In the waters around northwestern France, sightings of octopus used to be rare. Today, they're appearing in such great numbers that fisher men have begun trapping them from the same ocean floor that used to be fertile lobster grounds and are selling their new catch to seafood vendors in increasingly large volumes. One seafood auction house in Finistère, a region at the western tip of Brittany, reported that local fishermen had landed 1,200 tonnes of octopus in 2022. That was nearly three times the volume caught the previous year, and forty times more than in 2020.

By their nature, octopus don't want to work very hard to catch their prey. They look for food that can't escape quickly, and so shellfish are an easy target. Few meals offer less work than a trap full of lobster that can't escape. The octopus just has to follow them into the trap and start eating.

"In terms of food, the octopus is quite opportunistic: It feeds on what is easiest to catch," Dominique Barthelemy, a curator at Océanopolis, a science centre dedicated to the oceans, told the French magazine *Geo*. "If the octopus had a motto, it would be 'eat as much as possible.'"

French fishermen are being forced to shift quickly to this new and disruptive reality in their local waters. For now, most octopus caught here is shipped to Mediterranean countries, where it's long had a place in the cuisine. But if this voracious predator continues to push out France's beloved lobster, perhaps Restaurant Guy Savoy will need to find more space for it on the menu.

21

—

"I'm Still Here"

JONESPORT, Maine – To find the oldest lobster dealer in America, you just need to head down to the wharf in Jonesport, Maine, and ask for Sid. Inside a beige-and-red building, Bert Sidney sits perched in his cluttered office guarded by a friendly French bulldog, surrounded by pictures of his grandkids and a sign that says "Talk is cheap. Until you hire a lawyer."

"I'm still here, like a damned fool," he says, then breaks into a broad smile.

Few people know the state's lobstering history better than Sid, a kind-eyed, broad-shouldered grandfather now in his eighties, and few have seen more changes. He's the fifth generation in his family to run Look Lobster, which has been buying and selling lobster from Maine's fishermen since 1910.

In Jonesport, time seems to travel a little more slowly. In this sleepy fishing village that peaked at about two thousand residents, pickup trucks are the clear vehicle of choice and the Fourth of July beauty pageant is still the highlight of summer. It has a rich history in the fishery, and developed as an important commercial port and trading centre for the area's outlying islands. In the late nineteenth century, the Underwood fish packing plant on the waterfront was the largest of its kind in the

U.S. In 1871, it was paying sixty-two cents per hundred pounds of lobster. By 1901, Jonesport was home to the largest fishing fleet in the state.

In that era, Jonesport was home to a major shipyard, producing sailing schooners from Maine's abundant woodlands that would sail to the Caribbean with holds full of dried salted cod and return with rum and molasses. Sid's grandfather owned shares in the mercantile company that ran this well-worn trade route, and he operated a buying station on Port-aux-Basques, Newfoundland, where he would pick up salted fish.

When Sid was a young man, the Maine fishery was diversified, and fishermen earned a modest living year-round by harvesting different species throughout the seasons. But overfishing brought an end to that. In 1960, the last sardine cannery here closed, leaving lobster as the only reliable source of income. Jonesport became home to the largest lobster fleet in Maine. If fishing sets the rhythm of the community, lobstering is the music.

"Lobster is 80 per cent of our economy," Sid tells me. "And I'm worried the lyrics are about to change."

That's why the rising temperature of the water outside his office has him so concerned. Sid says it's affecting mortality rates and preventing lobsters from hardening their shells before they travel. They're more stressed when they're caught, and less able to withstand the long voyage to Europe or Asia.

His company began shipping live lobster overseas in the 1980s, mostly to Germany, France, and Italy. At that time, the lobsters had to be trucked three hundred miles to Boston's international airport overnight, and the shellfish cargo would often be bumped by federal mail that took priority. The trucks would have to make the long trip back to Jonesport, and many lobsters wouldn't survive that extra time in the truck.

"We'd get bumped every now and then, and they'd say 'Come and get these lobsters,'" Sid says. "If they had to sit on the damn truck for an extra day, they'd start dying. Whenever there were any problems, I'd see it in my paycheque."

Yet despite many advances in modern logistics, refrigeration, and sea-food export methods, the mortality rate among Maine lobster destined for overseas markets is actually increasing again. Rising water temperatures in the Gulf of Maine are causing lobster to shed their shell and grow a new one twice in a year instead of just once—a new phenomenon that often leaves their shells too soft to withstand the harsh demands of international travel. It's a problem that's increasingly worrying exporters who have built up a lucrative market of overseas buyers.

"The water temperature when he's in his natural habitat is too warm," Sid says. "Say that lobster is going to Rome. That's a forty-hour trip from A to B, but he's too stressed now. The lobsters aren't fully filled out, shell-wise, to make that trip. The shells on them are too *punky*. They're too flexible."

Sid wonders if the restocking programs that began in the Gulf of Maine in the 1960s didn't compound the problem. When he was a young man, preparing bait buckets for older fishermen on the docks at Jonesport, lobster caught off the coast of Maine and southwestern Nova Scotia had a more greenish tinge than the spotted black-and-brown versions pulled from the water today. He thinks the hundreds of thousands of northern lobsters brought in from the Gulf of St. Lawrence interbred with local crustaceans and began to change the population.

"I remember the first time I saw a shipment of lobsters from up there, and I was surprised at how speckled their tails were," he says. "Now you look at lobsters from New Brunswick, Nova Scotia, Maine, and they're speckled the whole way through. We brought in a whole different species as brood stock. We don't see any true green lobsters anymore."

The warming waters of the Gulf of Maine, the most productive lobster ground in America, has been something fishermen and biologists have been watching with concern for decades. Water here has been warming faster than most other parts of the ocean. The state's Department of Marine Resources has been tracking water temperatures in Boothbay Harbor since 1905, when they averaged around 7 Celsius.

In recent years, that same water has been measured closer to 11 degrees, averaged over summer and winter months.

A slight warming of ocean water actually helps lobster development in colder regions, allowing them to develop more quickly beyond the larval stage, when they are defenceless against prey. That's part of the reason Maine began enjoying record-setting harvests after 1990. But there's a tipping point. As ocean temperatures move beyond that optimal range, these incredibly sensitive creatures face serious problems. Lobster larvae struggle to develop into adults, and if they do, they're much more susceptible to shell disease and they struggle to moult.

Biologists generally agree that once ocean temperatures rise beyond 18 Celsius, even for a few days, the stress on American lobster becomes problematic and begins to do physiological damage. The past few summers, the water outside Sid's window has passed beyond that threshold with alarming regularity. During particularly hot stretches, the water in nearby Chandler Bay rose above 21 degrees—warm enough to begin killing lobsters. At that temperature, some lobster pounds had to shut off pulling water from the ocean into their lobster tanks because it was too hot.

"If they hit the bay with those warm temperatures, they'd had to close them off. It smothers them," Sid says. "They just couldn't stand it."

Sid seems to know everyone in the lobster industry, and has heard the reports about lobster disappearing from places like Buzzards Bay, Cape Cod, and Long Island Sound farther south as the waters there have warmed. He's watching declining catches. And he's worried.

"I can see that happening here, too," he says.

Biologists have been warning for a while that the future of the lobster fishery is worrisome. According to one study, in the *Proceedings of the National Academy of Sciences*, the lobster population in the Gulf of Maine could fall by 40 to 62 per cent by 2030. Those models don't even include other stressors of climate change, such as increased acidification and deoxygenation of ocean water, which could further reduce lobster numbers in the future.

For young lobstermen, heavily invested in bigger boats and needing bigger catches, pulling back on their fishing efforts risks putting them in a difficult financial position. Many, with too much overhead and rising debt, simply can't afford to catch less.

"The more I grow my operation, the more vulnerable I'm going to be," Elijah Brice, a nineteen-year-old lobsterman from Eastport, Maine, told *Craftsmanship Quarterly* magazine in 2023. "The more I invest in it, the more lobster I'm going to have to catch every single year to turn a profit."

Climate change isn't just reducing catches for fishermen in Maine and beyond. It's also producing more devastating storms and tidal surges that are destroying fishing infrastructure at an alarming rate. In January 2024, a record 14.57-foot-high tide was measured in Portland, Maine—the highest since measurements began in 1912—washing away fishing shacks that had stood at the edge of the ocean for more than a century. In Milbridge, just a short boat ride from Jonesport, the same storm tore away bait houses and shredded wooden wharfs. There was damage up and down the state's coast, raising concerns that fishermen and lobster companies wouldn't be able to rebuild in time for the opening of the spring season.

Storms have always been a fact of life for east coast fishing communities, but their increasing frequency and severity has both fishermen and climate scientists worried about the future. When Hurricane Fiona tore through Atlantic Canada in 2022, it did hundreds of millions of dollars in damage to the region's fishing sector, destroying wharfs, swallowing up lobster gear, and tossing boats onto land like toys.

The idea that Maine's billion-dollar lobster industry is facing a shrinking, less reliable future would have seemed unthinkable just a few years ago. Maine is synonymous with lobster—it has built a thriving tourism industry and countless souvenir shops around the iconic image of a red lobster. It has even put the shellfish on the state's licence plate. In towns all along the Maine coast, visitors line up around the block for lobster rolls, a sandwich arguably associated with this place more than anywhere else in the world.

But Maine has been through this before. Sardine canneries once dotted the state's coast, jutting off small-town shorelines and employing thousands of Mainers for more than a century. At the sardine industry's height, more than fifty canneries were in operation. By the early 2000s the industry was dying, and the last Maine cannery closed in 2010. Other fisheries have come and gone before.

Today, Sid knows the great lobster boom is fading. And he recalls a conversation he had in 1980 that warned him of all this. That summer, he was flying into Portland, Oregon, when the pilot flew close to Mount St. Helens, which had erupted just a few months earlier. That's when the man sitting beside him, a federal government scientist, began talking about dust clouds and catastrophic events that can begin to alter the natural world. He told Sid there would come a day when Maine would be warming and even snowmen would be rare. Sid politely listened. But secretly he thought the man was crazy.

"And I'm happy to get the hell off the plane," Sid says. "I could not believe it. He told me that, 'Hell, you ain't gonna have any more snowmen.' Yeah, he said that. He said you're going to have warmer weather in Maine."

Forty-five years later, he realizes how right that man on the plane was.

"I thought he was nuts. I couldn't get off the plane fast enough. But now, all these years later? How friggin' smart was he? He was *right*."

22

Lobster on a Roll

SAN FRANCISCO, California – It's nearly nine on a Sunday night, but the cars keep coming, unloading an endless line of passengers eager for a taste of one of the most well-travelled sandwiches in the world. They come by shuttle bus, taxi, and in their own vehicles and plop themselves down in the red metal chairs at picnic tables that surround the New England Lobster Market and Eatery, a warehouse-styled casual eatery just a short drive from the San Francisco International Airport.

The decor is a nod to the lobster shacks that dot the East Coast. There's the requisite rowboat and lobster traps on the ceiling and colourful buoys lined up on the walls. It's a deliberate copy of the small-town seafood restaurants found around New England, but no one is here for the decorations, anyway. The restaurant's main draw is the Maine-style lobster roll, which consists of cooked and diced lobster meat tossed in butter or mayonnaise and nestled in a grilled bun. It's all yours for $39.95 U.S., and comes complete with a cup of coleslaw and a handful of thick-cut potato chips.

The main ingredient in a true East Coast lobster roll—*lobster*—can't be sourced from anywhere other than from New England or Atlantic Canada. It's not like opening a taqueria in London or a sushi joint in Boston. Unlike shrimp, which is farmed around the world, or crab, whose multiple regional varieties are caught commercially, every one of those lobsters had to be plucked from North Atlantic waters, ferried to shore, trucked to a packing facility, and carefully loaded onto a plane for the time-sensitive flight across the country. To remind their customers of this unique terroir, New England Lobster has a large painted map of the East Coast lobster fishing region on their wall, showing the fishing ports they buy from, stretching from Prince Edward Island down to Cape Cod. The processing plant they partner with is marked, too, in southeast New Brunswick.

If someone had drawn up a map of lobster suppliers thirty-five years ago, it would have looked very different. When Marc Worrall, the co-founder of New England Lobster, started flying lobster from Boston into San Francisco, he sourced lobsters much farther south down the American coast, even as far as North Carolina.

"When we first started back in the late '80s, we were buying lobsters out of Rhode Island, then Long Island Sound during the month of May and June," he tells me. "And then slowly through the course of our time in business, we don't buy lobsters out of New York anymore, and no lobsters out of Rhode Island. I mean, even some lobsters came out of North Carolina back then. But that fishery, it's all gone."

Another change he's noticed in nearly four decades of buying lobsters is the decrease in size. It was common in the 1980s to get jumbo lobsters included in those shipments. Marc never sees lobster that big anymore.

"We would often take in lobsters from Nova Scotia and we would get twenty-, twenty-five-pound lobsters in the spring," he says. "Now, if we get a ten-pound lobster, we're lucky. So that tells you those lobsters are being caught before they can get any bigger, right? So as time goes on, you're going to see that lobster go down from ten pounds, then eight pounds, and just keep getting smaller."

Marc and his business partner, San Diego State college buddy Dave Collins, started the lobster roll venture as a way to make more money off "culls," or one-clawed lobster, which have poor value on the live market. They decided to cook these lower-value lobster themselves and turn the meat into sandwiches. It was an instant hit with West Coast foodies.

"Before that, lobster out here was a white-table, ninety-dollar meal," says Marc. "We wanted to change that and make it more casual."

Born and raised in San Francisco, the first time Marc saw a live lobster was in his mid-twenties, when he and his business partner began flying them in from Boston. Back then they had the energy to outwork everyone else. Sometimes working seven days a week, they just kept growing their business and expanding their orders from East Coast suppliers. But over time, Marc has developed a keen awareness that his key ingredient is undergoing significant changes and can't be guaranteed to last forever. As the lobster population continues to move north, he's had to move along with it, finding new, more northern suppliers. And he's competing with buyers from around the world in a way he didn't have to when he started.

"My real concern now," he says, "is the resource of lobsters. You know, there being so many lobsters being caught every day going to China, all over the world, where China didn't exist ten to fifteen years ago as a market. My concern is how can the ocean keep up with that demand? Back in the late '80s, or even in the '90s, you still had people just buying lobsters with a pickup truck and selling them for maybe twenty-five cents over their cost. Now, with what's going on with the lobsters, that doesn't exist anymore. I just don't see the ocean keeping up with the demand, and with the rising temperatures of the ocean."

The lobster roll is no longer a seasonal, low-cost sandwich found only in the Canadian and U.S. northeast. Companies like New England Lobster and Luke's Lobster, a Maine-based company that in 2023 had twenty-two locations in the U.S. and nine in Singapore and Japan, are feeding a global demand for an easier-to-eat, and easier-on-the-wallet, version of lobster. While people still line up for hours to enjoy the rolls at some of Maine's most famous lobster shacks, such as Red's Eats in

Wiscasset or McLoons Lobster Shack in South Thomaston, there are few corners of the world the humble sandwich hasn't reached.

Now you can get a lobster roll in Dubai or Mumbai or Berlin, if you so desire. Ironically, it was a French restaurant chain—Homer Lobster—that won the title of world's best lobster roll at a championship competition in Portland, Maine, in 2018. The sight of a twenty-four-year-old Frenchman, Moïse Sfez, taking home first prize in a lobster roll competition in the heart of U.S. lobster country says a lot about how global this American street food has become. Moïse called his version Le Connecticut, a nod to the sandwich's origins, and made it with Breton lobster, warm lemon butter, and a secret mixture of herbs, served in a toasted and buttered brioche bun. He told the Paris newspaper *Le Figaro* that he was introduced to lobster rolls while travelling in the U.S. as a teenager, fell in love with "the palace version of street food," and began making his own after training at a number of French restaurants and hotels.

Luke Holden, a former Maine lobsterman and Wall Street investment banker who cofounded Luke's Lobster, thinks there's still a lot of room to grow as he's expanding lobster roll shops to more corners of the globe.

"I think we have a lot more market that we've got to continue to develop. And there's a lot more headroom to continue to get one of the best products in the world to really great customers," he told me.

Luke's Lobster was born out of his frustration that in cities like New York it was extremely difficult to find an authentic lobster roll like those in Maine. Luke and his business partner Ben Conniff, both recent college grads, opened a two-hundred-square-foot hole-in-the-wall lobster shack in New York City's East Village in 2009. By 2018, it had grown into a $30-million company. In 2024, Luke's boasted $50 million in annual sales.

"Back when we started," says Ben, the company's president and a former food writer, "everything was extremely pricey, and appeared to be just kind of bathing in mayonnaise and celery and other fillers, served over a white tablecloth, a fine-dining experience. Nothing like the traditional lobster shack that you would experience in Maine. So that's what we wanted to create."

As they expanded to Asian markets, Luke's had to figure out how to source lobster, cook it, and nitrogen-freeze it for the long journey to the other side of the world via ocean freighters. The American company used a Japanese partner to manage their arrival into this new market, and it was immediately a hit. When their first location opened in Tokyo, customers lined up around the block to experience this new food, and the long lines didn't let up for years.

"Our partner understood that folks in Tokyo had a hunger for things that were popular in the U.S. that didn't exist at that point in Tokyo," Ben says. "So they leaned into that in a big way, like, 'Hey, you all love great seafood, but the one that you don't really have access to is Maine lobster, and you've never had a Maine lobster roll before.'"

Ben and Luke acknowledge that the volatile nature of lobster has made for some boom-and-bust years that they've had to develop their business around. They designed a vertically integrated model that gives them a buffer against those swings and allows them to ride out spikes in price. The company now processes its own lobster meat, and gets 60 per cent of its revenue from outside of its restaurants by selling products derived from lesser-value parts of the animal, like the shell, blood, and leg meat. A lot of that leg meat ends up in their branded lobster mac and cheese and lobster ravioli, which the company sells in grocery stores. The prized tails are sold to higher-end buyers. They even sell make-your-own lobster roll kits.

"We have a super-simple model and sometimes that works for us. Sometimes it works against us," Luke tells me. "We still have years where the cost runs up so aggressively, and we can't control that. But given the vertical integration, and the ability to kind of push margins around, we've always sort of prioritized a relatively stable lobster roll price at the restaurants, then move costs around to the other products that we produce."

The first year they opened, 2009, they were paying $14 a pound for lobster meat and selling a lobster roll (using a quarter pound of lobster) for about $14. In 2021, when the price of lobster spiked, they were paying $40 a pound and had to raise the price of their famous roll to $25.

As the price of lobster has climbed, Luke's has diversified into cheaper alternatives for its seafood sandwiches such as crabmeat and shrimp. The challenge for the future will be convincing people that lobster remains a sustainable, if pricey, choice. Even if it does come from the other side of the world.

"I think people are going to continue to become more discerning about the sustainability of the food that they eat," Ben says. "And so more people are going to be drawn to lobster, and lobster rolls, by the fact that it is climate friendly and that it is sustainably fished."

23

The End of McLobster

MONCTON, New Brunswick – Danny Moore is not a lobster man. He prefers a good steak, and on occasions when he does have seafood, he likes fish, battered and fried. But Danny, a Moncton businessman who owns three McDonald's franchises, has helped introduce thousands of people to one of the most iconic symbols of lobster's great boom—the McLobster sandwich.

By the early 1990s, lobster catches were becoming so plentiful, and prices still low enough, that restaurant chains and frozen food processors around North America began finding new ways to add it to their menus. It was an age of cheap, abundant lobster in the supermarket, spawning an invasion of frozen lobster ravioli, lobster pot pie, and breaded lobster bites into people's freezers and onto their dinner tables.

There was a time in this golden era when you could walk into a McDonald's on Canada's East Coast and pay just $3.99 for the fast-food chain's take on a traditional lobster roll. The original McLobster—or McHomard in francophone regions—was intended to be simple, using ingredients the restaurants already had on hand: a hotdog bun, shredded lettuce, and sauce from the McBLT, along with two ounces of thawed lobster meat, all wrapped in a foil bag.

The idea for a low-cost, mass-produced lobster roll was spawned by Danny's father, Gerry Moore, a former pro hockey player who owned McDonald's franchises in Amherst, just across the border in Nova Scotia, and in Sackville, New Brunswick, a leafy university town. He leaned on his contacts in the lobster industry from his years as an executive for a seafood company in St. George, New Brunswick, gathered a group of franchise owners from around the region, ordered in a bunch of lobster, and the McLobster sandwich was born. It was sold as a healthier alternative to hamburgers, something that could be eaten cold, without the need for frying or heavy sauces.

"People were talking about what we could offer to be different," Danny recalls. "At the time, a lot of people, especially seniors, were concerned about fast food. Lobster was seen as a healthy option."

The rise of the McLobster in Canada and, for a shorter time, in the U.S. came at a time of unprecedented catches in the fishery. In Maine in the early 1990s, lobster was still just $2.20 a pound when fishermen there began breaking harvest records that had stood for over a century. Lobster was suddenly everywhere, and companies rushed to find new ways to incorporate this other white meat into their offerings. It was the first wave in a campaign to make lobster a mass-market staple, an effort that also gave birth to lobster poutine, lobster burritos, and lobster macaroni and cheese.

"The root cause of lobster's slow migration from the white tablecloth to the drive-thru is that it simply isn't the scarce commodity that it once was," J.B. MacKinnon wrote in *The New Yorker* in 2015.

Part of that ubiquity was being caused by fishermen themselves. As catches rose in the 1990s, fishermen put more and more traps into the ocean, effectively feeding lobster with baitfish in a giant underwater farming operation. It's estimated that some 90 per cent of lobster that enter a trap are able to eat the bait and escape before they're pulled to the surface. With three million lobster traps in use in Maine, that's roughly 128,000 tons of bait delivered to lobsters every year in the state's inshore waters alone—"equal to one and a half billion Filet-O-Fish patties," MacKinnon wrote. Fishermen had to import bait from as far away as Japan and Portugal.

Back in 1992, the McLobster sandwich was his father's idea, but it was Danny who did most of the dealing with lobster wholesalers who could supply enough precooked, flash-frozen shellfish to feed the country's largest restaurant chain. He was also the guy franchise owners went to with technical questions if their employees were having a hard time preparing the sandwiches and thawing the meat without waste. Not every franchise manager loved it. Some found the cold sandwich too complicated to prepare and serve in a fast-food setting.

McDonald's was always a family affair for the Moores. Danny was pulled into the business while still in university, when his father called him and told him to get in his car and drive the two hours to Amherst, where he'd been offered a chance to open a McDonald's. Soon, the Moores' two-car garage was converted into a warehouse for the restaurant's supplies—piled to the ceiling with ketchup and sauce packets, napkins, Big Mac containers, wrappers, and other dry goods.

Gerry was an ideas man, a gregarious, personable owner who was happiest out in the front of the restaurant, meeting customers. Danny, who used to cook for his siblings and was more at home in the kitchen, had a head for numbers and was brought in as a partner.

"I don't know if Dad even knew how to use a microwave," Danny says. "He would come up with these big schemes and dreams, and Mom and I would have to sit down and figure it out."

Danny had some experience with difficult ventures for McDonald's. He'd been one of a group of restaurant managers sent to Russia in 1990 as part of the chain's earliest forays into the former Communist state, helping to open the first McDonald's in Moscow's Pushkin Square. The company had to start from scratch, building a supply chain to service their massive restaurant, which was the largest in the world, and an army of six hundred employees pumping out cheeseburgers, milkshakes, and fries.

As the McDonald's managers stumbled bleary-eyed onto an old Soviet Union–era bus, they were stunned at their first looks behind the Iron Curtain. And they were a surprising sight to the Russians, too, many of whom had never seen Westerners, clad in the neon ski jackets and other bright clothes of the early 1990s.

"Everything you saw around you was either grey or beige or brown," Danny recalls. "It was like stepping back in time. People looked at us like we were beamed in from another planet."

The campaign to expand into Russia began after the Montreal Olympics in 1976, when McDonald's Canada CEO George Cohon lent a stranded Russian delegation a corporate bus. Cohon saw an opportunity to get inside the Iron Curtain and hoped to expand into the Soviet Union in time for the 1980 Olympics in Moscow. But after three years of exhausting negotiations, the contract was killed after the U.S. and Canada boycotted the Games in protest of the Soviet invasion of Afghanistan.

It was another six years before Cohon began to make headway with Soviet President Mikhail Gorbachev. Two more years of negotiating, and Cohon finally had his deal. McDonald's opened on Pushkin Square in January 1990, with television networks watching the whole thing. By the end of the day, the restaurant had served thirty thousand hamburgers—more than any other franchise opening in the chain's thirty-year history. Russians were paying half a day's wages just to try a Big Mac meal.

"People were literally lined up for days," Danny recalls. "I've never seen anything like it. They were just mesmerized."

The reaction to the McLobster wasn't quite on that scale, but it was immediately a hit the first summer it appeared on the menu on Canada's

East Coast. In a little over two months, Danny and his family sold seventeen thousand sandwiches out of their Amherst location. Tour buses began stopping at the restaurant at the request of visitors who wanted to try the novelty sandwich.

"We'd see the buses pull in and we'd be shouting, 'Lobster! Lobster! Lobster!' We'd have a whole team of people in the back making lobster sandwiches," he says.

Impressed with sales, the executives at McDonald's Canadian headquarters in Toronto wanted to add the sandwich to their national menu, but felt they could improve on the recipe, Danny says. They hired a chef and began tinkering, adding green onions and celery and toasting the bun. The company switched the mayonnaise and fought with their lobster supplier on price.

"It kept changing and changing, but we'd already done the damage to ourselves," Danny says. "We were our own worst enemy because we kept changing it."

As the price of lobster rose, theft became another problem. McDonald's stores would buy large bags of shelled lobster meat, a tempting sight for some employees who began walking off with them. Some owners found that keeping the valuable lobster meat on-site was too much of a risk, and they began dropping the time-consuming sandwiches from the menu.

"It was an expensive inventory item," Danny explains. "Once you take it out of the freezer, you had two days to use it, and it took a day just to thaw it out."

Eventually, the wholesale cost of lobster became prohibitive for a fast-food environment, Danny says. By 2014, the company was paying $21.49 for a pound of shelled lobster meat—more than five times the price when the McLobster debuted two decades earlier, despite dramatically rising catches. Food costs for a McDonald's sandwich should be around 30 per cent. With lobster, it was more than half. As the menu price rose and rose and the novelty wore off, customers were ordering fewer and fewer McLobsters. The chain finally pulled the plug on the dream of cheap, fast, and readily available lobster sandwiches served up from coast to coast.

"People don't want to pay ten dollars for a sandwich at McDonald's," Danny says. "Even if it's lobster."

PART FOUR

BOOM AND BUST

24

Boom Times

COX'S COVE, Newfoundland and Labrador – If you want to measure the impact of lobster in Cox's Cove, a cluster of a few dozen homes wedged between the mountains and ocean in western Newfoundland, just look at the snowmobiles. In the 1990s, fisherman Rick Crane spent his teen years tearing through the woods beyond the cove on an "old beater" Ski-Doo, which his grandfather kept running with used parts, duct tape, and a backyard mechanic's improvisation. His family, like most here, was too poor to take it to a repair shop. How times have changed.

"Now, every family has two brand-new snowmobiles parked in the yard and a brand-new pickup truck nearby. We're eating *steak* instead of bologna," says Rick, referring to the fried thick-sliced bologna that has long been a staple of the Newfoundland diet.

Thanks to lobster, good times have finally returned to Cox's Cove, a village of 660 people not unlike a lot of coastal communities scattered along Newfoundland's western coast. While the lobster fishery is in decline in more southern regions, in Newfoundland, some four hundred miles north of the Gulf of Maine, the future looks bright. Catches have grown exponentially in the past ten years, and could double again within a decade, according to researchers at the Marine Institute, a Memorial University school that studies the maritime sector.

Fishermen in Newfoundland and Labrador landed 8,800 tonnes of lobster in 2024, compared with just 1,892 tonnes little over a decade earlier. More than $128 million worth of lobster was exported from the province that year, most of it to the U.S., but also increasingly to such places as South Korea and Denmark. Seafood companies are betting heavily on a growing lobster trade, spending millions on new holding facilities and improving airfreight infrastructure so Newfoundland lobster can more easily be sold around the world.

Lobster has quickly become one of the most important fisheries in the province, particularly on the island's west coast, at a time when shrimp and crab quotas are declining and the ghost of the cod moratorium still lingers. Newfoundland is enjoying a lobster population boom thanks partly to blind geographic luck. Its coastal waters, traditionally too cold to support lobster growth, are warming up enough to become productive fishing grounds. And while that trend line may be concerning decades from now, fishermen here are hauling in a growing harvest not unlike what their counterparts in Maine and Nova Scotia witnessed in the 1990s.

"Science was right, and the old fishermen, the dinosaurs, were wrong," Rick says, in his thick Newfoundland brogue that turns "father" into "fadder." "We're just in the right place at the right time."

Ironically, Newfoundland's lobster fishermen may also be benefiting from the collapse of northern cod stocks, which devastated coastal fishing villages in the 1980s and '90s. Codfish are a top predator of juvenile lobster, and their disappearance from the ecosystem in large numbers, combined with warming ocean waters, has given lobster the chance to flourish.

That was never the intention when the Canadian government shut down the entire cod fishery in 1992, ending five centuries of fishing off Newfoundland's Grand Banks. Fearing that the once plentiful cod stocks would dwindle to extinction, regulators felt they had no choice but to impose an open-ended moratorium. More than thirty years later, the fishery has still not reopened to commercial fleets.

The commercial extinction of northern cod bore out what people had been warning about since the 1960s—that marine resources once thought to be limitless were in fact vulnerable to overexploitation, and existing regulatory systems weren't strong enough to protect cod stocks. The moratorium was evidence that rules intended to safeguard cod populations hadn't kept pace with increasingly efficient fishing technology that allowed vessels to find and harvest unprecedented amounts of cod and allowed international fleets to work on the Grand Banks for months at a time.

For decades, governments often set cod quotas based on economic factors rather than ecological ones. Federal officials responsible for regulating the fishery consistently overestimated the size of cod stocks and, as a consequence, also overestimated the amount of cod fishermen could harvest at a supposedly sustainable level.

The cod moratorium crashed across coastal Newfoundland like a tidal wave, altering the province's economic and demographic makeup in ways still felt today. Almost overnight, more than thirty thousand Newfoundlanders were out of the job. Fish plants that had kept sea going communities alive were suddenly idled. Fishermen sold their boats. Families were torn apart as young men left in search of work. Tens of thousands departed remote outport villages, leaving isolated coves and bays that were the heart of Newfoundland's heritage and heading for factory jobs in Ontario and the oil fields of Alberta. For many years afterwards, the fishery was considered a dead-end industry. Rick grew up in the wake of the moratorium, when many Newfoundland families lived just above poverty levels.

"The only thing you could do to prevent starvation was fish in the spring and do construction in the fall and winter," he says. "We didn't have much money growing up. But I never needed anything I didn't have."

Rick was raised by his grandfather, Ludrick Crane, a lifelong fisherman who taught him how to work on the water. There was little in the way of other employment on the west coast of Newfoundland at that time, and many eked out a paltry living fishing and doing odd jobs the rest of the year. Canada's federal employment insurance program, heavily relied on by seasonal workers in coastal communities, filled in the rest.

"You just had to suffer it out and save your money," he says. "We had no luxuries back then."

Rick relied on his entrepreneurial streak to help pay bills. At age twelve, he was cutting loads of birch for firewood and selling it to his neighbours after school. As a sixteen-year-old kid, he'd buy a pack of cigarettes, then resell them individually at a markup to his classmates. He was more interested in his sales than his classes.

"School wasn't for me. I thought I already knew everything," he says.

Rick quit high school at seventeen and started fishing full-time alongside his grandfather Ludrick and grandmother Theresa Crane, but fought against a life at sea. They encouraged him to go back to school, get an education, and try for a government job. Like many young Newfoundlanders, he eventually moved out to work in Alberta's oil fields. But Cox's Cove was never far from his mind.

"My grandfather was my God. He was everything to me," Rick says.

When he moved back to Newfoundland in 2013, the banks turned him down for a loan to buy his first boat. So his grandparents signed over their modest home as collateral to help him get the vessel, which he named the *Corrina Maria*. Today, he fishes out of the *Crane's Legacy*, with his long-time friends Steven Park and Keith Cox.

When Rick's grandfather was a young man, conservation was far from mind and catches remained small. The fishery was essentially the Wild West until 1976, when limited-entry licensing policy was implemented and trap numbers were finally regulated. After 1998, the federal government further reduced the number of licences and increased minimum catch sizes, measures that may also be contributing to the healthy lobster population today.

Cox's Cove, a cluster of small vinyl-sided homes that all face the water, has been built around the sea since it was settled in the 1840s as a fishing and logging community. The fish plant is still the largest employer in the town. The cove looks much as it would have decades ago—piles of spruce firewood stacked high for winter form dividing lines between tidy yards crowded with lobster traps, snowmobiles, and all-terrain vehicles. The main street is still lined with fish stages, the single-storey sheds used for splitting herring, and rows of small wooden wharfs to launch boats.

Shielded from the North Atlantic by towering walls of rock and spruce trees, the Bay of Islands has become lucrative lobster grounds for fishermen who a generation ago had to venture farther offshore to catch fish. Each fall, Rick steams out toward the same grounds his grandfather fished, guiding *Crane's Legacy* through the grey North Atlantic and almost incessant winds with a cigarette clenched between his lips. Each plastic crate of lobster he fills is worth $500, and he's been filling a lot of them lately.

Newfoundland's new lobster boom is reminiscent of the peak reported more than 130 years ago, when the island's fishermen hauled in a record eight thousand tonnes, and that's with spotty recordkeeping that left many catches off the books. This first lobster boom spawned a network of canneries that dotted the island's coast and sent tins of Newfoundland lobster around the world. But a lack of regulation, allowing a free-for-all with no restrictions on minimum size or egg-bearing females, led to a stock collapse in the 1920s that didn't rebound for decades.

Canneries and lobster pounds sprung up across Newfoundland during the last lobster boom in the late 1800s, supplying packers that shipped lobsters to Pictou, Nova Scotia, or directly to Boston. Often the lobsters were crated with ice and transported in sailing ships, or later, in fast diesel-powered former rumrunners that were bought after

Prohibition. It was a surprisingly efficient system for the time, and worked well enough for an island far removed from customers on the mainland. As lobster catches began to grow again after 2010, exporters began exploring ways to charter cargo jets to get live lobsters to overseas markets.

Record lobster landings have also launched a spending spree that has trickled down to Newfoundland's boatyards, car dealerships, and yes, even snowmobiles. In the spring of 2023, Rick opened the first live lobster holding tank owned by a fisherman in the province. His facility, which cost him $500,000 to build and with space for twenty-two tonnes of the crustacean, allows him to hold on to his lobster when the price dips and sell when it improves.

In 1999, when he first started fishing, a good season—two or three months—was around one and a half tonnes of lobster. By 2006, three tonnes was the norm for fishermen in Cox's Cove. In 2024, some of the most productive fishermen along this rocky coast were hauling that in a week. When Rick was a teenager, lobstering was difficult work, with little money to be made. It was not a job many wanted to do. He'd freeze his fingers banding lobsters on the back of the boat, hauling every trap up by hand because his family couldn't afford a motorized hauler.

Catches were always small back then, he recalls. Within a few weeks of the season opening, it was normal to catch maybe eighteen or twenty lobsters in a day's work. They'd save them up all week to sell just a crate or two for about $300, if the prices were good.

"It was just survival after the first few weeks," he says.

That all began to change around 2015, when catches began to take off and the warming waters off western Newfoundland proved to be ideal breeding grounds for juvenile lobster. Rick thinks there will be decades of good catches ahead. Newfoundland is just lucky to be situated in a northern region where the lobster seem to be headed.

"I don't see me running out of lobster in my lifetime, and I don't see it in my son's life either, to be honest," he says. "Fishermen who don't see that, who don't think global warming is true, they need to get out from under a rock. Reality is reality, and we're getting the lobster that they're losing in Maine and Nova Scotia."

Of course, climate change is not only bringing more lobster to once colder regions. It's also bringing more violent storms, which can wreak havoc on fishing communities. In September 2022, Newfoundland's southwestern coast was battered by Hurricane Fiona, a deadly storm that washed away dozens of homes in one of the worst natural disasters in the province's history. More than 130 harbours used by fishing vessels around the region were damaged by the powerful winds and storm surges, along with hundreds of millions of dollars in lost or damaged boats and fishing gear.

But in Cox's Cove, spared the worst of Fiona because of the large granite mountains that protect it from the Atlantic, there's a lot of optimism for the future of the fishery.

"Finally," Rick says, "the poor bottom-of-the-barrel lobster fishermen of western Newfoundland, who had nothing else, are getting to enjoy a few luxuries of life."

25

The Town That Lobster Saved

TIGNISH, Prince Edward Island – Stuck way out on the flat north-western tip of Prince Edward Island, far off the beaten path for most tourists, Tignish might be just another fading village in Canada's Maritime provinces if it weren't for lobster.

Many small rural communities in this region have long struggled with a shrinking, aging population, few jobs, and even fewer young families. Tignish, a sleepy, tree-lined community whose population peaked in the 1970s and which lost its railway service a decade later, could have easily been one of those towns. Instead, its schools are bursting at the seams, its renovated hockey rink is full of children, and there's a thriving Filipino community that has brought new life to a village where most families traditionally had Acadian, Scottish, or Irish roots.

The Filipinos are here because of Royal Star Foods, a processing plant and seafood co-op that sold $145 million worth of shellfish in 2022, most of it lobster. There's so much lobster coming into the plant that the company had to bring in hundreds of temporary foreign workers because there weren't enough locals to work in the place.

"Without these people, we wouldn't be able to run the plant," says Francis Morrissey, Royal Star's affable manager, who talks in his cluttered office in a pair of black jogging pants, his green rubber boots kicked off. "We'd have to shut down because we simply couldn't get local workers. We'd advertise for six months and we couldn't find any locals to take these jobs."

More than three hundred Filipinos have become permanent residents, buying up empty homes, starting families of their own, and sparking a mini construction boom. Their place in the local economy can't be over-stated, Francis says.

"There's no houses with the lights turned off anymore. There's no houses for rent anymore, because they've all been rented."

Nearly seventy Filipino families have bought homes in the community of about 720 people. The company has helped them get settled here by sponsoring relatives who want to come and work and assisting them in applying for residency status with the government. There are nights Francis goes down to the plant to help his employees pack lobster and he realizes he's the only one in the room with a Canadian passport. They've created a Filipino volleyball league, and have brought diversity to the local high school, which fifteen years ago would have been almost exclusively white.

The lobster boom's impact can be seen everywhere here, from the busy marine mechanics' shops to the apartment buildings under construction on the edge of town. Maritimers are unflashy by nature, but they display their newfound affluence in shiny new pickup trucks parked in the driveway, stacks of pristine lobster traps and polished million-dollar fishing boats on stilts nearby. Brand-new homes, and recently renovated ones, dot the coastline on the scenic drive out to the North Cape lighthouse, which Francis's grandfather used to operate in the Second World War with a kerosene lamp.

The story of lobster in fishing communities along the Gulf of St. Lawrence has been one of unprecedented and rapid growth. Landings of the prized shellfish nearly quadrupled in most fishing zones along the gulf between 2006 and 2021, according to the federal Department of Fisheries and Oceans. When Francis, now sixty-two, began lobster fishing with his father, a good season was just over a tonne. Today, the boats rumbling in and out of the narrow channel near Tignish bring in as much as twenty tonnes in a year.

"The fishing has been strong," Francis says, in an understatement. "As the fishery has boomed, everything else has boomed right along with it."

He attributes the recent growth in the lobster population to the collapse of the groundfish sector—namely, cod—in the early 1990s. Fishermen have also adopted more sustainable methods, such as increasing size restrictions, using cleaner-burning diesel engines, and switching to traps that disintegrate over time to allow lobsters to escape if the traps are lost.

Climate change may also be playing a role, the former fisherman says, though he's quick to add he's not an expert. But there's no denying the arrival of right whales in the gulf, a relatively new phenomenon as the massive creatures and their enormous appetites have followed zooplankton into cooler waters. Zooplankton, those tiny microorganisms that include crustaceans, water insect larvae, and aquatic mites, are also a favourite food of young lobster, he points out.

Francis's company is also enjoying these boom times, but he's not one to show off his success. His pickup truck, a 2011 GMC Sierra, is parked at an angle outside the co-op's front doors. It's got 475,000 kilometres on it, and more dents than the pockmarked dirt roads that crisscross the province. He keeps the keys in the ignition, in case anyone needs to borrow it.

"Pretty much everyone on P.E.I. has bumped into my car at some point," he jokes.

Francis remembers the hard days when lobster catches were slim here and many families had to supplement their incomes by harvesting Irish moss, a seaweed used as a digestive aid and health food. He spent a summer hauling the seaweed off the rocks and drying it so he could buy his first bicycle. Most families in that era were large and poor, and they lacked creature comforts like indoor plumbing and central heating.

Many homes had an outhouse, and were drafty and cold at night, he says. Perhaps that's why he brings a practical-minded approach to his job, choosing to eat his lunch with his employees and rejecting the trappings of most CEOs.

"I never wear a suit or a tie. And I ain't putting one on, even when I die. I want to be buried in a T-shirt. If they put me in a suit, I'm coming back to raise hell," he says. "I think one of the worst things a person can do is think they're better than someone else."

Francis refuses to let his success go to his head. There's no doubt that with historically high catches and a seemingly unending global demand that has turned lobster into a luxury food, Tignish is riding a new wave of prosperity. Francis points out that all the lobster fishermen have nicer trucks than he does. But even though landings are substantially larger than they were a generation ago, fishermen can't catch the lobster fast enough.

"There's not enough lobster in Canada for all the millionaires in China who want it," Francis says.

Part of Royal Star's success is also due to Francis's ambitious expansion efforts, including adding a 1.5-million-tonne-capacity cold storage facility contained inside a series of nondescript beige warehouses along the wharf—the largest like it in Canada. He also invested in a brine-freezing facility that can hold up to 8 million tonnes of frozen lobster, which allows the company to hold on to lobster when the price is low. Thanks to investments like these, Royal Star has grown from a regional company to a firm that sells lobster around the world, most notably to the U.S., Europe, and Asia. Francis still laughs when he describes how he watched Chinese teenagers pose for endless pictures with Canadian lobsters, cooked in front of them at an Alibaba retail store.

As Tignish has grown more international, so has the market for its lobster. Increasingly, live P.E.I. lobsters are being flown to the Middle East, where countries such as Saudi Arabia are developing a taste for the crustacean. In Qatar, the British-owned Burger & Lobster restaurant chain will serve you a steamed Atlantic lobster plucked and flown in from Canadian waters for the equivalent of $70 Canadian. For $80, you can have a lobster grilled with chermoula paste, a traditional punchy North African marinade.

The global demand for Atlantic lobster was not something the founders of the Tignish Fisheries Co-operative Association could have imagined when they created the co-op that Francis runs today. Formed in 1923 to stop local processing plants from exploiting fishermen, the union demanded that its members sell exclusively to them, in exchange for a fair, reliable price. It was the first co-op of its kind in Canada. Prior to organizing, many fishermen were living at poverty level—always in debt to the fish plants, and paid in credit that could be spent only at the company store. Many were trying to dig their way out of this cycle by fishing "half line" agreements, in which the processing plants got half of their catch free of charge because fishermen had to rent their boats from the companies themselves. In those early years, companies often paid a flat rate by the season, no matter the amount of fish caught. It was commonplace in that era for the banks to automatically refuse credit to fishermen.

The cooperative association in Tignish put a stop to these exploitive practices and helped spawn a larger social movement in the community, first with a credit union and then, in 1938, the Co-op Store. Both businesses are still there today, anchoring the village's small main street, right next to the giant Christmas tree made of old lobster traps, erected as a memorial to fishermen lost at sea.

The cooperative movement that spread across the eastern provinces owes much of its origins to these early fishermen's unions. One of the Tignish co-op's founders, lawyer and businessman Chester McCarthy, went on to become the first president of the group of co-operatives known as the United Maritime Fishermen. This group and others like it helped change the fortunes of fishermen.

The Tignish fishermen's cooperative is also one of the last surviving co-ops like it in the industry, and diversifying into live lobster when most stuck to processing is a big reason. Francis was the first to begin supplying P.E.I.'s famous lobster supper restaurants with live local lobster at a time when most suppliers were bringing theirs in from the U.S.

Today, 235 fishermen own the processing plant and share in the proceeds of Royal Star's profits each year. Francis says there's plenty of lobster off P.E.I.'s shores, and plenty of work ahead for his plant.

"I think we have a bright future," he says. "As long as we can continue to find workers."

26

—

Well-Travelled Lobster

BOSTON, Massachusetts – In 1896, the Underwood name was synonymous with canned food in America. But William Lyman Underwood, the grandson of the man who pioneered the canning industry in the U.S., had a serious problem with one of his most popular products—lobster.

By the end of the nineteenth century, canneries had sprung up all along the Atlantic coast, pumping out a seemingly endless supply of processed seafood for people who lived nowhere near the ocean. Hundreds of them dotted shorelines in Canada and the U.S., offering fishermen an efficient way to get their lobster processed and in the hands of customers long distances away.

The explosion of the canning industry brought North American lobster to a wider market, but it also brought about the first concerns about overfishing. Before the 1860s, there was virtually no research on or management of the sea-based fisheries. Fishermen caught as many lobster as they could, and the plants worked ceaselessly to process and sell what they brought in. But governments in both Canada and the U.S. soon learned there were limits to what the ocean could provide. "Now the rapid development of lobster canning on the Atlantic and salmon canning on the Pacific would change the picture. Plants were going up in any empty cove. It took only two decades for the lobster and salmon canning industries to go from zero to practically maximum production, bringing threats of overfishing and catch decline," writes Joseph Gough, a Canadian fisheries historian.

The Underwood company helped turn lobster into an international commodity for the first time, making the crustacean of choice for European aristocracy suddenly accessible and affordable for the middle and working classes. The packer was one of the first to make North American lobster a truly global food, shipping their preserved products, first in hermetically sealed glass jars and later in tins, to the East and West Indies, Hong Kong, India, Gibraltar, South America, and the Philippines.

Underwood became a household name in the early nineteenth century thanks to its Deviled Ham, originally sold in glass jars. Much of the company's early success was based on the work of French confectioner Nicolas Appert, who had spent years experimenting with ways to preserve soups, vegetables, dairy products, and more in glass jars, mostly by sealing them with cork and sealing wax before processing them in boiling water. In 1810, Appert won a prize equivalent to $65,000 Canadian today from Napoleon Bonaparte, who chose his method to keep foods for long periods to feed his army on its march to Moscow. Despite his discoveries, Appert died poor, but his legacy in modern-day food processing is unquestionable. He's why the process of heat-treating canned food is sometimes called appertization.

By the late 1800s, when Underwood and other factories were pumping out an endless line of canned lobster, catches of American lobster reached historic highs that wouldn't be matched for nearly another century. Unless you lived in New York or Boston or coastal New England, where smacks—boats with saltwater holds—could deliver cargoes of fresh lobster within the short window before they needed to be cooked, the only way most people typically enjoyed lobster was boiled and packed in a tin.

But the science of canning was still evolving. Consumers complained that the quality of tinned food was spotty. The world may have developed a taste for the crustacean, but controlling bacteria was still an evolving science, and spoilage was a common problem. The canning industry even had a term for this stomach-turning phenomenon: black lobster. In an 1896 paper for the Massachusetts Institute of Technology, Lyman Underwood wrote, "The contents of such cans were found to be badly decomposed, in some cases entirely liquefied, much darkened in color, and of a very disagreeable odor." Underwood was determined to figure out a way to improve the canning process to stop bacteria from destroying his products.

Helped by MIT food scientist Samuel Prescott, Underwood spent months in the lab in 1895 investigating what was causing his cans to spoil. What they identified were types of bacteria that formed heat-resistant spores that fuelled bacterial blooms. More importantly, they learned that these spores could be killed by canning at 121 degrees Celsius for ten minutes. Their new sterilization process would transform the science and technology of canning and usher in a world full of safe canned products. When it came to lobster it was a game changer, allowing a perishable seafood to travel farther and last longer than ever on customers' shelves.

For much of the nineteenth century, canning remained a laborious, by-hand process that involved cutting strips of metal and soldering together a can's sides, bottom, and top. That remained the dominant method for canning seafood for well over a century. By the 1940s, these once ubiquitous little factories were vanishing from the coast, although the seafood processing business remains an important part of the lobster fishery in the Gulf of St. Lawrence region. Increasingly, however, even fishermen in this northern part of the fishery have been voting in favour of phasing out catching the smaller, "canner" lobsters, for conservation reasons.

"It used to be there were canneries in almost every little port," says Bernie MacDonald, manager of the Ceilidh Fishermen's Co-op on the western shore of Cape Breton island. "Now they're all gone."

Canneries were eventually replaced by larger, modern processing facilities and the growth of the live-lobster trade, which allowed seafood companies to get a much better price for the shellfish. Initially this involved shipping lobster in live-well smacks and placing them on trains or trucks bound for Boston, then, as today, a major seafood distribution hub. The final step in the globalization of lobster came as exporters learned how to keep lobster alive for weeks or even months after they were caught, chilling and aerating them in indoor holding tanks until carefully packing them in cargo planes, all with the goal of keeping mortality to a manageable level.

Companies like Benson Lobster, a family-owned wholesaler on Grand Manan Island in New Brunswick, have played a significant role in that evolution. In the 1950s, Lloyd and George Benson began building outdoor open tidal pounds to store live lobster for customers around the Maritimes. These rustic holding pens worked well enough, but didn't allow for controlling water temperature or effectively keep predators out. In the 1970s, their sons and grandson formed Benson Lobster, moving lobster from their temporary pens by herring boats and later transport trucks that took the live loads down to Boston.

By 2009, the company switched to an indoor model, using tank houses that pull seawater from the bay and keep it chilled and circulating constantly. Lobsters are kept separated in little plastic boxes called condominiums; these are placed in large rectangular pools of bubbling, chilled seawater where the lobster essentially hibernate for as long as two or three months at a time. While they're dormant, Benson's staff are constantly monitoring oxygen and ammonia levels in the water, and keeping the temperature at 4 Celsius. This process, meant to mimic wintertime hibernation conditions under the sea, keeps lobster a lot healthier and ready for travel than the previous outdoor pounds.

"There's a lot of 'Greenpeace people' out there who want them crawling around on the bottom of the ocean, feeding them every three days, stuff like that," says Casey Benson, one of three sons now involved in the business. "But you can't keep them alive that way in storage because they'd eat each other. They're scavengers. They're basically vultures that just kill each other."

When it's finally time to ship the lobster, they're placed in thirty-pound boxes and loaded onto trucks bound for airports in Halifax, Toronto, New York, Chicago, Montreal, and Boston. From there, Benson works with freight forwarders to send them around the world. Benson Lobster began shipping directly to international customers only in 2019, and that side of the business has exploded ever since.

Casey's father, Morton Benson, just sits back and smiles when he's asked how much things have changed at Benson's. When he followed his father into the company as a young man, this was still a largely regional business, operating seasonally, and the fishermen who supplied them needed to hold down multiple jobs to pay their bills. The idea that they'd one day be shipping planeloads of live Grand Manan lobster to China would have seemed laughable. Even more unimaginable— there are now two Chinese-owned lobster exporters on the island, just down the road from them, competing for the same lobster.

"When I first started out, the furthest any of our lobster would go would be Boston or New York," Morton says. "It's nothing like today. Today, the volume we're doing on Grand Manan is probably five times what it was in the 1980s. It's a good thing Asia has come on board, too, because, my goodness, where would you go with five times the volume in North America? It had to go somewhere else or else it would be worth absolutely nothing."

In the nineteenth century, state and federal governments in North America began toying with the idea that if lobster couldn't survive transport over long distances, perhaps they could move the entire fishery itself.

In the early 1870s, America's population was nearly thirty-nine million, more than double what it was three decades earlier. Many of its rivers in the heavily populated East were being fished out, and there was worried discussion at upper levels of government about how to supply food to this growing young nation.

Livingston Stone, a Harvard-educated federal fish commissioner, saw the need to establish fisheries in parts of the U.S. where some food staples had never been seen before, including lobster. He began an ambitious program to transplant striped bass, shad, salmon, and catfish to the west coast, and the federal government hired him to open California's first freshwater fish station.

Livingston, a former church pastor who was also an accomplished chess player, had some initial success with transplanting non-native species to the other side of his rapidly growing country. But one of his most spectacular failures was an attempt, in 1874, to create a Pacific coast fishery for Atlantic lobster from New England.

Lobster by then was a wildly popular seafood in the U.S. Northeast. Plenty had tried before, and all had failed, to bring this crustacean from its natural habitat in New England and establish it in other regions. According to a history of these attempts compiled by the Smithsonian Environmental Research Center, one of the earliest attempts came in 1814, when the U.S. fisheries commissioner tried to extend the range of lobster southward by acclimatizing eggs in warmer water, ordering a box of live lobsters to be deposited near Charleston, South Carolina. In 1884, sixty-three live lobsters, including some egg-bearing females, were transplanted to Back River Light, a lighthouse near Norfolk, Virginia. In 1885, another hundred were transplanted to the shoreline of Chesapeake Bay. Similar attempts were made in British Columbia by Canadian fisheries officials.

In 1874, Livingston, then the deputy fish commissioner for the U.S., converted a Central Pacific Railroad car into a two-thousand-gallon aquarium on wheels in preparation for a trip to San Francisco. He used ice, milk cans, and a hand-pumped aeration system, and he loaded up this rail car with East Coast seafood staples, including shad, striped bass, catfish, salmon, eels, and lobsters. He hired an assistant, M.L. Perrin, to pluck 150 live lobsters from a seafood distributor in Boston and collect casks of seawater from outside Boston Harbor, where it was cleaner. But almost as soon as the train left, problems began. Just one day into the trip, the lobster were already dying. Perrin blamed a lid falling off one of the containers.

In his account of the voyage, republished in *Fishermen's Voice* magazine in 2011, Perrin wrote: "Whether the falling of the lid was the cause of their death, we could not quite decide; but it seemed very probable that it was because the air pumped into the tank after the lid fell, having no means of escape at the top of the tank, exerted a great pressure upon the water and in this way killed them, and also because of the impure air which was confined inside for some time without being replaced by purer."

What Perrin, and other marine biologists of his era, didn't understand was the impact of disease among lobster confined to small places. A small percentage of all wild lobster have gaffkemia, a blood disease caused by bacteria. When an infected lobster dies, the scavenger nature of lobsters leads others to tear the animal apart. In a confined tank, this can spread millions of live bacterial cells into the water, quickly infecting the other healthy lobsters.

Perrin's experiment had another problem—maintaining an optimal water temperature when transporting live lobster. In the mid-1870s, when refrigeration technology was still in its infancy, the temperature of the ocean water in the bin would have risen as the train made its way inland. (The first U.S. patent for a mechanically refrigerated railcar wasn't registered until 1880.) Previous attempts to use ice had limited success, but only in colder weather, never for a cross-country trip with a creature as sensitive to temperature as lobster. Perrin tried to control the temperature by stacking blocks of ice on top of the wooden crates, and sponging seawater over the lobsters packed on straw in the crates, but he couldn't slow the dying. "After the fifth day," he reported, "crowds of lobsters take offense at something, and revenge themselves by dying.

The reason of their death was wrapt in mystery. Numerous experiments always failed to bring any regular results, and nothing certain could be gleaned from them. Theorizing about lobsters' chances of life is vain when applied in practice."

Perrin soon learned lobster become stressed once they're out of water. "Every time the car-doors were opened or the atmosphere around the lobster-boxes disturbed, there inevitably rushed upon them a draught of warm and dry but injurious air, fatal at once to a lobster in case the current strikes it," he noted.

After a week, with his stock of live lobster rapidly depleting, Perrin gave two "very healthy and active lobsters" to a fisheries superintendent in Utah, who placed them in the Great Salt Lake. As Perrin's specially built railcar left Utah that night and continued westward, what remained of its lobster cargo was a pitiful sight. Just eight animals had survived the trip to that point. There remained only one pail of salt water. Perrin sent a telegram ahead, and a freight train met the expedition the next day with four barrels of Pacific Ocean water near Beowawe, Nevada, a pass-through trading post where the mountains seem to open up like a gate to the valley beyond.

Of the 150 lobsters that Perrin left Boston with, only four made it all the way to San Francisco Bay. They were carefully placed in the Pacific Ocean at the Oakland Long Wharf nine days after they had been taken from the Atlantic Ocean. Perrin complained afterwards that it would have been better if state fisheries officials had ordered them to be put farther out to sea, where the water was not so warm, and more salty. He was blunt in assessing the lobsters' chances of survival in their new environment: "The four lobsters themselves probably did not live," he wrote in his report on the failed trip.

Still, fisheries commissions continued to try for many years afterwards to establish lobster colonies, attempting transplants in Texas, England, and Japan. None of these attempts took into account how vulnerable American lobster are to changes in their environment, and how they've evolved over millions of years to thrive exclusively in the cold waters off New England and Atlantic Canada. These experiments showed just what incredibly sensitive creatures lobster are. Too much or too little oxygen or sunlight, fluctuations in heat or cold, and crowding can all lead to stress, disease, or death.

As late as the 1960s, the Canadian government was still entertaining the idea that a commercial lobster fishery could be created in Pacific waters using lobsters transplanted from the east coast. The scientists proceeded, hoping to succeed where at least fourteen previous attempts could not over the past century. Despite that poor track record, there was a surprising confidence—even worry about a potential "population explosion" if the lobster thrived in their new environment. Government officials, long accustomed to domesticating countless livestock species, had assumed in their arrogance the same could be done with lobster. They were wrong. R.J. Ghelardi, a federal fisheries scientist, wrote in the opening passages of a 1966 report: "In part these changes [to food production] were brought about by man's spectacular success in domesticating and transplanting wild grasses, grains and animals; but introductions or transplants have not always ended successfully."

The Canadian efforts in the 1960s learned from those past experiments, taking steps to isolate the lobsters on their trip across the country to reduce the spread of disease. In another case, the scientists chose Fatty Basin, a shallow inlet on the west coast of Vancouver Island, for their experiment; they felt its sheltered waters and rocky bottom most closely mirrored lobster's Atlantic Canadian habitat.

One hundred and eighty-four lobsters, all weighing over a pound each, were caught in the Bay of Fundy and taken to a marine research station in St. Andrews, a tourist village on the south shore of New Brunswick. Only ninety-six survived a week in quarantine, before they were flown to Vancouver Island. There, they were kept in cages at the bottom of the basin and fed a diet of beef liver and herring. As the transplants died, they were replaced with healthy ones being kept alive at a facility in Nanaimo. The experiment, while it showed that lobster could live in the waters of Fatty Basin from June through November and that they moulted and grew in roughly the same way they would have in the Atlantic, was ultimately called off due to lack of funding and limited success.

The dream of creating a Pacific coast fishery for the beloved *Homarus americanus*, which had taunted biologists for over a century, was officially dead.

27

Fishermen Farmers

SPRUCE HEAD, Maine – With its sun-bleached fish houses, Cape Islander–style boats bobbing in rocky harbours, and wooden docks stained black by the tides, Maine's Midcoast region looks like it came straight out of a Norman Rockwell painting. And yet even here, in the heart of America's lobster fishery, people are thinking about life after lobster.

For Bob Baines, who has been fishing these waters since 1982, that means building intricate webs of rope suspended seven feet underwater by heavy anchors. As lobster catches have declined along this once plentiful coast, he and other American fishermen have begun farming kelp, a large brown seaweed that grows in dense groupings. It's become a leafy insurance of sorts against an increasingly uncertain future.

"I was sixty-something years old, and it was like, 'Do I really want to do something different?' But if I wanted to keep lobstering, I needed to diversify," he admits during a chat at his kitchen table while his new puppy chews away at his hand, demanding attention. "People ask me, 'Do you consider yourself a fisherman or a farmer?' Oh no, I'm still a fisherman. But this is three-dimensional farming, or at least that's what I call it."

Bob uses his knowledge as an experienced fisherman to run his kelp farm. It all began when, at the suggestion of his daughter, he obtained a lease from the state on a protected bay and began building his four-acre farm, and doubled that in 2023, enough to produce around 100,000 pounds of kelp a year. Each November he "plants" his crop along an underwater thousand-foot-long rectangular structure made up of moorings, buoys, and seeded ropes that crisscross back and forth. Working with two other men, one on the deck of his fishing boat the *Thrasher*, another in a smaller skiff nearby, it takes only a few days to install the ropes and anchors each fall. In Bob's view, it's just practical, giving him protection against declining lobster catches and some supplemental income to help pay bills.

Bob is at the end of his career as a fisherman after witnessing the great boom of the late 1990s and early 2000s and then the near decade of declining catches that followed. Lobstermen in the state caught more than sixty thousand tonnes in 2016, twice what they'd caught a decade earlier, and nearly four times what they landed in 1996. But the declines since have been alarming and sudden, leaving fishermen like Bob very worried about the future.

"Some guys just want to ignore this," he says. "But the writing is on the wall."

Kelp's prime growing season is through the winter, when most lobstermen aren't out fishing anyway, so it just works well, Bob tells me. The springtime lobster fishery in Maine isn't what it used to be, so it's easier, he says, for fishermen now to wait until their kelp is harvested before they return to the sea with their traps.

Like Bob, an increasing number of lobstermen in Maine are partnering with companies such as Atlantic Sea Farms, which processes kelp into a range of products, including toothpastes, shampoos, salad dressings, puddings, cakes, dairy products, frozen foods, and pharmaceuticals. A typical kelp farm can bring in between $20,000 and $57,000 in extra income for a fisherman. Some are doing it on a larger scale, and earning more growing kelp than they ever did with lobster.

No one needs to tell Bob that the Gulf of Maine is warming, rapidly changing the conditions that not long ago made for prime lobster fishing. He's watched it happen over the course of his life on the water.

"Climate change is real," he says. "I've seen it. When I first started fishing in the wintertime, you had to break ice to get out to your boats. And we don't see that anymore."

But it's not just anecdotal evidence that has him convinced. Bob also works with a researcher at the Woods Hole Oceanographic Institution in Falmouth, Massachusetts, who has been measuring Maine's water temperatures with probes placed inside lobster traps. He's one of about forty fishermen who participate in the study, which uploads temperature readings in real time via their cell phones. The research shows that the Gulf of Maine is heating up faster than 99 per cent of the world's oceans.

"You can just see the increase in the water temperature," Bob says. "You can argue what's causing it, but you can't argue with the numbers."

Warming ocean waters can't be blamed exclusively for the declines in lobster numbers that fishermen are now seeing off the coast of Maine, Bob says. The seas here have been fished so heavily and so relentlessly that lobster stocks have had no chance to recover. Bob has witnessed this change happen since he entered the fishery, when fishermen were still using wooden traps and smaller inshore boats. In Maine, there are now three million lobster traps in use, roughly twice the number that were in the same water in the 1980s, and fishermen are going farther and farther away from shore to find new lobster grounds.

"We have so much fishing effort now," he says, referring to what biologists also call fishing intensity, a measure of the amount of fishing equipment and activity in a specific area. "In this state, we pound them. Summer, fall, even winter, we catch them all, you know, so there's not much left in the spring before they shed. It's a terrible model."

As the catches grew along the New England coast, the boats grew along with them. Fishermen began working their traps in heavier rotations, carrying as many as two hundred traps at a time. Setting that many traps at once demands a bigger deck, and so boatbuilders began increasing the size of their vessels. In the 1980s, a forty-foot boat was unusually large. Today, it's considered small, with most new boats approaching fifty feet. Fifty years earlier, fishermen here still used fifteen-foot boats, often powered with old automobile engines repurposed from local scrap yards. It was almost subsistence fishing in that era; there was little extra money to put into their business.

For nearly a century, lobster catches in the state of Maine remained stunningly reliable. It was as steady as the daily tides, good enough to build a consistent, if meagre, source of income for many coastal economies. Then the boom started. By 2012, Maine's lobstermen were catching six times what they had caught in the 1980s. It made for a frenzied time in the fishery, launching a spending spree and a flood of investment into the industry.

The boom had been decades in the making. In the late 1960s, the centre point of the U.S. lobster population was somewhere near New Jersey. By 1998, it had moved north of Boston. By 2010, it was off northern Penobscot Bay, Maine. As the population moved northeast, in search of cooler water, catches in Maine exploded.

Yet when it goes, it goes quickly, as fishermen south of Cape Cod can confirm.

In the 1950s, a typical lobsterman on the Maine coast might have operated 150 traps. By the 1980s, growing competition was driving some of them to use as many as 500. Today, many fishermen in the state are running 800 traps. This marked increase in effort makes little sense, argues Jim Acheson, the well-respected economic anthropologist at the University of Maine who spent much of his career studying fishing communities in coastal Maine. He points out that there's a fixed number of lobsters in any given area that grow to legal size each season. If lobster populations are unchanging, all lobstermen are doing by increasing the number of traps is increasing the cost of their gas, bait, and gear, and working harder and harder—but they're not catching more lobster. At times, as warming waters created a short-lived spike in the population before rising beyond ideal conditions. At times, overfishing has also resulted in a glut of lobsters that caused the price fishermen get for their catch to drop in half. Now that catches are declining, many fishermen are overcapitalized for an era of peak lobster that no longer exists and can't afford to scale back. "This escalation makes no sense from an economic point of view," Jim wrote in his book *The Lobster Gangs of Maine*. "This escalation first came about in every harbor as one or two men began to build more traps to obtain a larger portion of the available lobsters, and others increased their own traps to keep up. . . . Although most lobstermen understand [the problem this causes] they have steadily added more traps to keep up with the competition."

Bob, who is still tall and sturdy in his late sixties, says he "fell into fishing" as a young man and never found anything better. Unlike a lot of Maine lobstermen, fishing wasn't his family business; his father was an auditor for the Port Authority in New York. But after his father retired and moved the Baines family to rural Maine, he started working for the Spruce Head fishermen's cooperative, an organization that his son would eventually lead.

Stories like Bob's, of Maine fishermen who spent their entire lives catching lobster and are now trying to diversify, are everywhere. They know they can no longer depend on the natural wild fisheries to keep on performing year in, year out.

"My nephew went back and got his captain's licence for the coast guard," Bob tells me. "And so the last six or seven years during the winter, he goes on a tugboat. So, you know, up and down the coast, you have a number of different fishermen doing those sorts of things. We have to."

Many others parked their boats and started growing oysters, a common pivot for fishermen farther south. One of them, a sixth-generation fisherman named Joe Young, has been increasingly keeping his boat anchored while he puts on waders and wanders down into the salt pond behind his house. Joe, who lives in the scenic fishing village of Corea, Maine, now tends to tens of thousands of tiny oysters growing inside porous boxes, stacked up like underwater file drawers, instead of plopping his lobster traps at sea.

When *The New York Times*'s Jess Bidgood went to visit him in 2017, he showed her photographs taken by his late aunt, Louise Young, of his village in the 1940s, '50s, and '60s. In those days, the fishing boats were loaded with cod, pollock, and herring as well as lobster.

"Look at what you see here in the photos, and how much of that's gone," he said then. "What if 50 years from now, what we have now is gone? And it's oysters and mussels and kelp?"

South of the Canada-U.S. border, the end of the lobster boom has become one of the most pressing issues for an industry confronted by a rapidly shifting economic landscape.

"Climate change really helped us for the last 20 years," lobsterman Dave Cousens told *The New York Times* in 2018. "Climate change is going to kill us, in probably the next 30."

In 2012, Maine's fishermen realized just how risky their overreliance on one species was. That season, nearly fifty-five thousand tonnes of lobster were caught in Maine's waters, nearly six times more than what the catch was in the 1980s. The ensuing glut of lobster caused the price to drop from around four dollars a pound to around two dollars. To combat falling prices, lobstermen chose to fish even harder and haul in more than before, only compounding the problem.

Sam Belknap, director of the Island Institute's Center for Marine Economy, in Rockland, Maine, recalls, "That year, we had an anomalously warm winter and spring, which caused early in-migration and shed of lobsters from offshore, and they flooded the market prior to the Canadian processors being ready. And then the supply was up, and the demand was down, so it caused a price collapse. Bologna was more expensive than lobster that summer. I think that opened up a lot of folks' eyes."

Sam, the Ph.D.-educated son of a lobsterman, says this shock caused a lot of fishermen to begin looking elsewhere for income, such as kelp farming, and caused families to discourage their children from following them into the fishery. But with so many people overcapitalized for so-called peak lobster, the journey back to more sustainable lobster harvests will be painful.

"We're starting to see some of those telltale indications of the longer-term impacts," Sam says. "And while it's going to take us a while to reach those levels that were the historic average of around thirty million pounds, that journey back is going to be very difficult because the fishery is so structurally different than it was in the mid-'90s and the preceding generations."

While older fishermen, operating smaller boats and fewer traps in the older fishing style, may be able to weather that storm of shrinking catches, those who spent big to go further offshore can't afford any downturn in landings.

"If your business model is based on a million-dollar loan for your fifty-foot boat and new gear every year fishing offshore, and the catch starts to decline in any significant amount, you're pretty vulnerable," Sam says. "That's because you're operating at a commodity level at that point, where you're really focusing on volume rather than value. So those larger fishing vessels fishing offshore, they're going to be the first ones to feel the kind of true economic impact of this."

Sam also worries that a shrinking catch may lead to more conflict on the water as fishermen begin to move around and enter territory they haven't fished before. But convincing people to consider alternatives, and to catch at a more sustainable level, can sometimes be a tough sell in an industry where some lobstermen are raking in as much as $200,000 a year and high school students are taking home as much as $60,000 a summer working as a deckhand. For many coastal communities, lobster is basically the last refuge of the marine economy.

"It's going to take some structural changes, it's going to take less focus on kind of out-baiting your neighbour's traps, and focusing on effort reduction," Sam says. "You can maintain a viable fishery, even as we start to see these declines, climate-driven and otherwise. But the question is, are folks going to choose to go down that path? Or are they going to insist and fight to the last moment, 'Well, I'm doing things the way they have traditionally done it,' even though their tradition in many cases only extends back ten or fifteen years, so they don't understand how the fishery operated in the past?"

The time is now for Canadian and American governments to begin preparing for the end of peak lobster and help coastal communities transition to a more sustainable future, Sam argues. He warns that if Maine's coastal communities don't begin shifting to a post-lobster-boom economy, they could be headed for an economic collapse not unlike that seen in America's coal country or in the hollowing out of the Midwestern states during the dust bowl of the 1930s.

If that happens, more of Maine's coastline could begin to look like many of the now abandoned islands that dot the state's shoreline. With no other commercial-scale marine species able to take the economic place of lobster, an entire way of life could disappear. In the 1950s, more than three hundred islands in the state had year-round communities built around fishing, much of it the now defunct cod and herring fisheries. Today, only fifty year-round island communities remain, a troubling trend that is part of the reason for the creation of the Island Institute, which has worked to protect Maine's working waterfronts while expanding its reliance on aquaculture.

"If things go as poorly as they could for the lobster fishery because of climate change, if the impacts exacerbate, if the regulations don't allow for fishermen to do things in a new way in a timely manner, you're going to start talking about coastwide economic collapse, which is scary," Sam says. "And if that happens, you lose the economic heart of a community. The community disappears or falls to a level of kind of barely subsisting."

28

Lobster in the Library

CUTLER, Maine – In the spring of 1986, Brian Beal was trying to help a group of fishermen in the far eastern tip of Maine start a lobster hatchery. So he went, naturally, to the library. He didn't go there to learn, of course—Brian was already a skilled aquaculture biologist. But he needed space to grow his lobster babies before they were released into the ocean. And it just so happened the little Louise Clements Memorial Library, which doubles as the small community's town hall, had some room in the basement. And, more importantly, it was next to the harbour, so a supply of fresh seawater would always be close at hand.

The village of Cutler, Maine, is a short boat ride across the international border from Canada. It's protected from the Gulf of Maine by the rocky peninsula of Western Head that forms Little Machias Bay and has long made it a natural setting to fish from. It's the kind of place where the harbour comes alive at three in the morning with the sound of marine diesel engines warming up for a day at sea, and the smell of bait is never far away. Lobster has long been an important part of the local fishery. But in the early 1980s, landings were declining, and fishermen wondered if a hatchery program could help them improve the local population. Across the whole state, the harvest had dipped below nine thousand tonnes, about a fifth of today's catch. The lobster boom that would bring unprecedented revenues into coastal communities was still several decades away.

Born in nearby Machias, Maine, in 1957, Brian grew up in a fishing family and was never far from the water. His grandfather taught him how to harvest lobster and clams, tie knots, and build his own traps. His summers were spent on his grandfather's boat. By the time he was finished high school, Brian had his own skiff and was managing about fifty traps of his own. He kept on fishing while in college, going out on the water after classes ended in the afternoons, but eventually became less interested in harvesting from the sea than understanding how the ocean's ecosystem worked. Following four years at the University of Maine, he pursued his master's degree in marine science. After three years in the swampy heat of North Carolina, he returned to his home on Maine's foggy, temperate coast. He started working as a biologist in the University of Maine's school of marine sciences, and eventually founded the Downeast Institute, a hatchery and research centre on Beals Island.

Brian had a question that many other biologists and entrepreneurs had tried to answer before: If humans can farm salmon, oysters, shrimp, mussels, and carp, why can't we do it with lobster? The village eagerly got behind the project, and two local research assistants joined Brian in creating one of the state's first community-run lobster hatcheries.

"We're a fishing community, so people were all for it. We had big hopes," recalls Cynthia Rowden, a former town officer whose grandfather used to haul traps near Cutler by hand, going from buoy to buoy using only sail power. "If the lobster were to leave, we'd have to curl up and go away, because we rely on our fishermen for everything."

The library allowed Brian and his team to install six one-hundred-gallon conical tanks that raised baby lobster during the larval stage until they were ready to be released into the ocean. A long pipe drew seawater up from the harbour, and it circulated through the tanks, while algae that was grown in an outside greenhouse was used as food for the larvae.

Each year, Brian would choose a handful of egg-bearing lobsters caught by fishermen and carefully cultivate those little black eggs into larvae in the library's basement. In the wild, a one-pound female lobster can carry around eight thousand eggs tucked on the underside of her tail, and those have a survival rate of about 1 per cent. Under Brian's watch, however, the survival rate improved to about 50 per cent.

It wasn't easy. The first year, the eggs kept dying at high rates, and Brian was spending a lot of time in a chemistry lab at the University of Maine testing out a series of crude experiments to improve their survival. Eventually he settled on a formula: As soon as the female lobster flipped her tail and prepared to release her eggs, you needed to be there to catch them with a net. The larvae, he learned, had about fourteen seconds to get into a food-enriched environment, or else they would start dying. It required near-constant monitoring.

"I mean, when the female flips its tail, even if it's one thirty in the morning, you need to be there," he says.

Once hatched, the larvae were immersed in water enriched with tiny brine shrimp, also known as sea monkeys. Brian bought his brine shrimp in dried egg form, hatched them in seawater, and fed them microscopic algae until they had grown big enough to become food for baby lobster. The lobster larvae were then vigorously aerated, which provided oxygen to the water and, more importantly, kept the larvae from coming into contact with each other, preventing cannibalism. The water was moving so much it almost looked like a pot of boiling seawater. This mimicked the ocean, where lobster larvae are carried off the bottom to the midwater range where food is more abundant and currents can whisk them away to new places, until they eventually grow big enough to settle on the ocean floor.

Lobster larvae in their earliest stages almost resemble mosquitoes swimming around in the water. By the time they grow into what biologists call stage 4 lobster, they have become bottom dwellers that begin to look like tiny lobster, about the size of a penny, with fine hairs that eventually fall off and are replaced with legs. At this stage, they're released back into the ocean, with the hope they can grow into adult crustaceans in a few years. They're awkward and often helpless against currents that pick them up and carry them around the ocean.

"It's kind of like if somebody strapped a piece of Styrofoam to your or my arms with some electric tape and threw us overboard and said, 'Okay, go ahead and dive down,'" says Brian, clad in a plaid shirt, glasses, and moustache. "But you wouldn't be able to do it, because of the buoyancy of the Styrofoam."

Between 1986 and 1992, Brian and his research assistants pumped out almost 100,000 lobster annually from their library hatchery using this method. How many survived to adulthood is unknown, but the experiments proved to him that, under controlled conditions, lobster larvae could be consistently raised into juvenile lobsters. In 1993 and 1994, he was invited to go to Ireland, where he demonstrated the same technique using European lobster.

Lobster hatcheries around the world have attempted similar experiments like this, usually funded by governments and fisheries groups concerned about shrinking lobster stocks. The first hatcheries began in New England and Atlantic Canada in the late nineteenth century, when concerns over declining catches spurred federal and local governments to build hatcheries along the coast that raised millions of lobster larvae and dumped them back into the ocean.

In St. Andrews, New Brunswick, a lobster-growing facility built in 1974 was able to produce juvenile lobster year-round. The St. Andrews Lobster Culture Laboratory, which showed that selective breeding could produce strains of distinctly coloured lobster, closed in 1983 when government priorities shifted and the site lost its funding.

Brian's experiments to raise lobster from larvae to juveniles kept running into the same problem, just like many before him—it takes too long to grow a lobster to market size, which is over a pound, a process that requires between five to seven years in the wild.

"All of them had to do with the same thing, which is, you know, how do we put a one-pound lobster that's cultured on the market?" Brian says. "And, quite frankly, I don't think that's the approach that I would take."

Instead, Brian argues that farmed lobster could have a market if they were sold as something similar to crawfish—a much smaller freshwater cousin of lobster that is already being grown in Southern states such as Louisiana. His experiments have shown that American lobster can reach crawfish size in a year. Selective breeding, like that studied by the lobster lab in St. Andrews, can also manipulate the colour of the mini lobster's shells, which he thinks would help set them apart from wild-caught lobster.

"I think a tiny blue lobster would look pretty good on a dinner plate in Tokyo," he says.

Brian settled on his mini-lobster method by experimenting with juvenile lobsters placed inside plastic soda bottles that he drilled holes through before submerging in the bay. The young shellfish grew larger than those set free into the bay.

But the experiments stopped when he couldn't find any funding to support further development of his lobster cultivation method. The joy of realizing it could be done, that lobsters could be grown from eggs into adults, eventually gave way to frustration that no one would fund him to pursue it further.

"That first moment, it's pure joy. It's pure joy until you find, you know, years later that nobody gives a shit," he says. "Then you think, 'All right, well, you get this stuff published. And maybe, you know, three hundred years after you die, someone will pick it up.'"

Plenty of others have tried to meet the challenge of farmed lobster before. In the 1970s, John Hughes, a pioneering scientist with the Massachusetts Lobster Hatchery and Research Station, told *The New York Times* he could grow market-sized lobster in as little as eighteen months—twice the amount of time it takes to grow a human, the newspaper pointed out. But John's claims that lobster's growth could be sped up dramatically by heating seawater to 21 Celsius were unsubstantiated, and he wasn't able to replicate his experiments at a commercial level. The project eventually lost government support.

Nevertheless, some continue to bet heavily on farmed lobster. A company called the Norwegian Lobster Farm claims it can grow "plate-sized" lobsters, between nine and twelve ounces in size, from egg to market in about twenty-four months. The company takes seawater used to cool the Green Mountain data centre in Finnøy, Norway—a massive network of computer servers used by companies such as TikTok to store and process their IT operations—and feeds it into a land-based aquaculture system. The water coming out of the data centre is 20 Celsius, which happens to be an optimal temperature for lobster growth. Each lobster is grown in its own container, to keep them from cannibalizing each other, and fed pellets by computer-controlled advanced robotics.

Rick Stein, a British TV personality and celebrity chef, endorsed a similar concept in 2008, backing a project by Swansea University's Centre for Sustainable Aquatic Research to grow lobster in Norway. But that experiment, aided by a £1-million investment by the European Union, ran into the same challenges that have plagued so many of these farmed lobster projects: It's an expensive and time-consuming way to grow a shellfish the ocean makes for free.

The Norwegians, accustomed to commercial-scale salmon farming since the 1970s, still believe in the technology and say they have a proven concept that's sold ten thousand of these mini lobsters to high-end restaurants. They also claim the model can be scaled up by cooperating with data centres around the world that use seawater to cool their operations, and currently own the largest brood stock facility in Europe. But there remains plenty of skepticism in the seafood industry that it can be scaled up to a commercially viable alternative to wild-caught lobster.

As wild catches decline, and the price of lobster rises, that skepticism may begin to dwindle.

"It remains very difficult," says Lin Li, who manages sales for a delegation of Norwegian seafood companies in China. "Because if you confine them together, lobster will try to eat each other. They need to be kept apart."

In 2012, Darden Restaurants, the American restaurant giant that owned Olive Garden, LongHorn Steakhouse, and Red Lobster, invested $300 million U.S. to develop the world's first integrated lobster aquaculture park in Malaysia. The project would grow tropical spiny lobster, not hard-shelled North American lobster, and it was aimed at providing the Red Lobster chain with a steady supply of one of its most expensive proteins. Back then, Darden was buying some 7 per cent of all the warm- and cold-water lobster harvested around the world each year. The twenty-three-thousand-acre lobster farm in Sabah, East Malaysia, on the north coast of the island of Borneo, was reported to have the capacity to produce forty million pounds of lobster by 2030. At the time, Red Lobster said it wanted to find a way to make the industry completely self-sustainable.

Darden Group had spent years, and millions, trying to solve a problem aquaculture companies have long been stumped by: How do you shorten the long growing period required for adult lobsters, while keeping them well fed and free from disease in close quarters? In 1998 the company began funding the Australian Research Council's work in developing lobster farming programs, where it took a whopping sixteen years before researchers were able to raise lobsters in captivity.

The Malaysia experiment was dropped when Darden divested all its interests in the project in 2015, after it sold off its Red Lobster brand. For now, Red Lobster says all lobster sold in its restaurants remains wild-caught. But the company's new owner, Thai Union Group, was a seafood processor that was heavily involved in aquaculture, and all the shrimp served in Red Lobster's "endless shrimp" promotions were farm-raised.

More than a decade later, the Malaysian government is still holding on to hope that commercially viable farmed lobster can become a reality. Many small businesses there have jumped on the bandwagon as the government promotes the fledgling industry in rural areas. Dozens of farmers are producing farmed freshwater lobster using aquaponic systems and selling on a smaller scale, while also supplying lobster larvae to other aquaculture operators. They keep their lobster in small ponds and feed them a wide-ranging diet including potatoes, carrots, and spinach. Lobster's ability to eat anything, unlike most species of farmed fish, is part of the appeal.

But the Malaysian efforts to grow lobster also have their detractors. Robin Zhou, a sales manager at QQ Supreme, a Malaysian aquaculture company that raises tiger prawns, dismisses the idea that farmed lobster can ever work in Malaysia's heat. He says it's too challenging to control water temperature in the vast outdoor ponds needed to farm the shellfish.

"It's too difficult for our climate," he says. "It's always summer there. It's too hot."

In North America, meanwhile, the argument against farmed lobster isn't a problem of technology; it's one of economics. The wild-caught variety is, for now, still abundant, even as the fishery is moving northward. And with so many fishermen heavily invested in catching lobster at sea, not on farms, there's little political will to consider alternatives.

Gregor Reid, director of the Centre for Marine Applied Research, a branch of Nova Scotia's food and aquaculture development agency, argues that a serious downturn is needed in North America's traditional lobster producing regions before governments and entrepreneurs will be willing to start investing in lobster farms. And if farmed lobster can be raised anywhere, in a controlled environment meant to mimic the once productive conditions of the Gulf of Maine, it may make more sense to do that in more cost-effective locations closer to the markets in Asia, he says. That means Vietnam, rather than, say, Yarmouth.

"I think it would just end up fairly low down the priority list," he says. "So I think for people to be receptive of aquaculture, it would have to be depending on where the people in the fishery start kind of moving on. But right now, with so many communities dependent on this and doing fairly well, it would be really tough."

That may begin to change as more southern regions of the fishery begin to look for ways to replace the jobs lost by declining wild lobster populations. With clever marketing, Gregor says, consumers just might embrace a smaller, less meaty version of lobster that was raised in captivity. The seafood business already understands how powerful rebranding can be. Patagonian toothfish became known as Chilean sea bass in 1977, thanks to a fish wholesaler named Lee Lantz who was looking for a way to make it more attractive to the American market. Similarly, New Zealand's aquaculture industry was able to create a demand for the unfortunately named slimehead fish just by changing its name to orange roughy and marketing it to international buyers as a new kind of fish.

While there remain challenging obstacles, such as lobster's propensity for spreading disease and their cannibalism when they're confined together in small spaces, and their slow growth rate, plenty of aquaculture experts argue that the science around farmed lobster has already been proven. It's just waiting there for us when we need it. Brian Beal thinks the North American lobster industry shouldn't wait until the fishery collapses in southern regions before taking aquaculture more seriously. As catches move northward, the economic impact will be staggering if there isn't something else to fall back on.

"But here's the deal," he says. "This is the perfect time to be trying this stuff out. You know, the industry isn't in jeopardy at this point. We think that, well, we *know* that water temperatures are increasing. And as we look at the Gulf of Maine, we see the ports of entry where lobster fishermen have been bringing their lobsters in for, you know, decades and half centuries, and things like that. And we see the ports where the most lobsters are being caught shifting towards the north. So this is exactly the time to be doing this."

29

Life After the Boom

GRAND MANAN, New Brunswick – It used to be that the real money on this island, a cluster of spruce trees, Pentecostal churches, and ocean-worn volcanic rocks, was on the beaches, so slick with wet seaweed you couldn't stand.

On Grand Manan, about a ninety-minute boat ride from the Canadian mainland at the edge of the U.S. border, people have long made a living harvesting dulse—that purple leaf that grows thick along the rocky shore and is picked at low tide. It's an important additive in health foods with its high concentrations of iron, vitamin C, and potassium and its antioxidant and anti-inflammatory properties. People have been collecting, drying, and eating dulse on Grand Manan since the era of the sailing schooners. Those wooden ships are long gone, and so are the sardine factories and smoked herring plants they used to supply. Plenty of fisheries have come and gone on this iron-bound island, too, but dulse has always endured.

Elton Greene, one of the most experienced dulsers on the island, knows Grand Manan's rugged shoreline by memory. As a boy, he'd often fall asleep curled up in the bow of his father's boat, soothed by the briny smell of the Bay of Fundy. Elton, who grew up spending his summers harvesting dulse on the beach with his siblings, is one of the few dedicated dulsers left here. Most mornings in the summer, he can still be found hauling his small dory up the rocky beach with its burlap sacks full of wet seaweed, which he drives to an empty lot near his house and spreads it out to dry in the sun.

Few people want to do this painstaking work anymore. There's better money to be made now as a deckhand on a lobster boat, a job that can bring in six figures in a season. Old-timers will tell you that lobstering used to be a meagre, part-time job. Not anymore. Things began to change in the early 2000s, when fishermen here began hauling in more lobsters than they'd ever seen before. While catches had been growing steadily for a few years, it felt as if overnight, untold hordes of lobster just showed up on the shoals around Grand Manan. Lobstermen were catching more in a day than their grandfathers would catch in a whole season.

"We didn't know where they were all coming from," says Elton, who calls his little boat *Zoso*, after his favourite band, Led Zeppelin. "There were suddenly lobster *everywhere*."

The great lobster boom changed everything about life on Grand Manan. The old stereotype of the poor fisherman earning a modest wage from the sea evaporated as islanders started buying new trucks, upgrading their boats with modern engines, and doing expensive renovations on their homes. Deckhands suddenly had money to burn and went looking for places to spend it. All that cash brought new social problems to an island where alcohol was long shunned by the church, and drugs flooded in. But almost as striking as how quickly the boom came is how quickly it's passing.

"There was a joke on Grand Manan that you can catch lobsters in your backyard, lobsters in the woods. They'd say, 'Yeah, there's too many lobsters,'" says Andrew Westgate, a marine biologist at the University of North Carolina Wilmington who studies lobster out of a research station in Grand Manan. "It used to be when the fishery opened in November, it was a grind for six straight weeks, fishing every day. But now that period of time has gone from six weeks to about two weeks. That slowdown happened much more quickly."

Overfishing, climate change, and market forces have pushed the fisheries in many east coast communities into an extreme reliance on lobster, to the exclusion of all other fisheries. Grand Manan is no different.

"There used to be a very diverse group of fishing activities," says Heather Koopman, the marine biologist who has led lobster tagging projects around Grand Manan. "And there was a seasonality to it. You went lobster fishing in the fall, you scalloped in the winter, then in the spring you'd do a little bit more lobster fishing. And in the summer, you put up your herring weir and you fished herring. There were these different fisheries that were supporting the island. And now it's almost exclusively lobster."

On Grand Manan these days, lobster drives the island's economy. Apart from some bed and breakfasts, a few salmon farms, and a handful of tourist attractions, little else generates employment here. On the ferry ride from the mainland, I tucked into a plate of fried clams, surrounded by truck drivers, construction workers, and contractors, almost all of them paid by the money generated by lobster. As lobster goes, so goes Grand Manan.

Brian Guptill is the president of the Grand Manan Fishermen's Association. Now sixty-five, he still has a youthful mop of blond hair on his head and big, broad hands. He's been fishing lobster since he was a teenager, when he was hired as a deckhand on his uncle's boat in 1982 after quitting college to become a marine diesel mechanic. In those days, if you caught five thousand pounds in a season, you were lucky. Today's commercial fishermen regularly haul in ten times that amount.

The next summer, Brian bought his own wooden boat and a lobster licence and went into business for himself. He fished with the rotating seasons, also harvesting scallops, groundfish, and herring, and took his lobster traps out of the water after Christmas. Like many islanders, he went into the woods in the winter to cut timber.

"The numbers just weren't there," he says. "There just wasn't enough in the traps to make it a full-time job."

For about two decades, he carried on like that, making just enough of a living to stay afloat and raise a family. Then fishermen began hauling in record catches, year after year. The island's only dentist felt this economic boom firsthand, Brian says.

"He knew things were taking off when people he'd given up sending bills to came in and started paying their bills."

Today, Brian pays his deckhands—the entry-level position on a lobster boat—over $100,000 to work for him. Some don't have a high school diploma. That kind of income was unimaginable for lobstermen when he was starting out in the 1980s.

When Brian was a boy, the poor kids brought lobster sandwiches to school. His grandparents would gather lobsters off the beach after storms and spread them on their fields for fertilizer. Today, most of the lobster caught off Grand Manan is treated like gold—a precious cargo that is shipped around the world to consumers willing to pay premium prices for it.

"Back when I started, years ago, there was just a few guys fishing around the island and on top of the shoals," says Brian. Today, fishermen scour nearly every square foot of ocean floor, trying to catch as many lobster as they can. Technological advances have allowed them to fish in areas they couldn't before. Bigger buoys, boats, and anchors, along with stronger mechanical hoists and a whole new generation of nautical aides, have also allowed them to catch far more lobster on each trip. And as the value of the total lobster catch has grown, more people have competed to get a piece of it.

In Maine, where eighteen thousand people depend on the lobster fishery for a job, the steady decline in catches since 2016 has fishermen bracing for an uncertain future. As they contend with a changing ocean environment and new rules designed to protect rare whales, the downward trajectory of the lobster harvest has plenty of people on edge, worried that a collapse like that seen in Rhode Island and Connecticut is headed their way. Some are hoping it's a temporary decline and cling to signs of optimism in models that suggest more baby lobster are settling in deeper, colder waters farther off the coast of Maine. But no one is pretending there isn't a pronounced, and painful, shift happening in a fishery that has been a part of North America's coastal communities for centuries.

As catches decline, it's expected fights over fishing grounds will continue. And not just over the growing First Nations fishery in Canadian waters and among commercial fishermen themselves, but fights between American and Canadian fishermen battling over the last few patches of seabed in contested areas along the border. There remain tensions over the so-called grey zone, the fishing in the Gulf of Maine area claimed by both countries, and some predict more violence if it isn't resolved.

"It needs to be shut down and turned into a conservation area. Someone is going to get hurt," says Howard Robbins, a fisherman from Lubec, Maine, who now fishes in Canada.

The portion of the harvest being brought in by Indigenous fishermen, meanwhile, grows every year. Brad Small, the captain who fishes on the New Brunswick mainland out of Dipper Harbour, wonders if the simplest solution is to divide up the total lobster catch, with hard limits for both sides—an idea that a few years ago might have seemed blasphemy.

"Maybe it needs to go to a quota system," he says. "Gives the Natives their share, and let us keep ours. Maybe that's the only fair way to do it."

Plenty of fishermen are worried about the sustainability of the North American lobster stock, but they don't agree that a quota system is the way forward. Many point out that a quota system was in place in both Canada and the U.S. when the northern cod fishery collapsed. Part of that failure can't be blamed on the quota system itself, of course, but mismanagement by government officials beginning in the 1950s that overestimated the size of the fish stocks. It exposed the problem of relying on commercial data, which suggested there were twice as many fish as the research data did, in setting harvesting targets while ignoring warnings about fish populations showing obvious signs of overfishing.

There are clearly things that can be done immediately to help improve the sustainability of the lobster fishery. While the U.S. has a maximum size restriction, protecting the most productive breeders, Canada has no such limit. If preserving the abundance of lobster as long as possible is the goal, there's no good reason why Canadian officials shouldn't impose a limit. And minimum size requirements play a critical role, too. U.S. regulators, concerned about warning signs in studies of juvenile lobster, planned to raise the minimum size that can be legally caught to 86 millimetres of "carapace" length—a measurement of the lobster's body minus the head and tail—by 2027, up from 82 millimetres in 2024.

Fishermen often grumble about these changes, fighting against what they see as restrictions on their catch, but in the long run it's usually in their best interest. Dave Cousens, former long-time president of the Maine Lobstermen's Association, made this point in the 1990s during protests over a plan by state regulators to increase the size of the escape vent on lobster traps, to allow smaller lobsters to slip out and grow a little longer.

"Fellas, we sell them by the pound. If they get out and get fatter, they ain't walking to Spain," he once told a crowd of fishermen in Portland, Maine, who were threatening to riot over the change, according to a 2018 account in the *Fishermen's Voice*.

In spite of the most well-intentioned management practices, the human race keeps growing larger every year and in the process consuming more seafood, including lobster. Removing demand, a critical piece in the story of lobster, from the equation is impossible. But allowing market forces alone to determine the fate of lobster is a formula doomed to fail. Increasingly, the future of seafood consumption is in domesticated fish, not wild species caught out on the open ocean like lobster.

That means that as lobster catches decline in North America but global demand stays high, as is widely expected, it will increasingly become a luxury food reserved for special occasions. In the winter of 2024, after a disappointing fishing season in New England, Nova Scotia, and New Brunswick, live lobster at Toronto's St. Lawrence Market was selling for thirty-five dollars a pound, a price that would have been unimaginable a few years earlier. The era of cheap lobster, once fed to servants in New England and served as a last resort by landowners in the colonies, is already a quaint footnote in the history of the shell-fish. In the future, we can only think of it as among the world's most expensive proteins.

Geoff Irvine, executive director of the Lobster Council of Canada, tells me, "We have to think about the world being a smaller place, and lobster as something going on someone's plate in a high-end restaurant. And it's also in retail stores all over the world. So it's really important to remember that, because we're competing with Wagyu beef. We're competing with king crab legs, we're competing with snow crab, we're competing with high-end beef."

But there is a tipping point. As the price creeps higher, it's thought that consumers and some restaurants will begin to choose lobster less often, easing some pressure on the price. It's just not clear where that ceiling is. Just when lobster roll shops in Maine predict the high-water mark for the price of their sandwiches, the price blows right past it.

Another challenge will be convincing people that eating this increasingly expensive protein is good for the environment. Already, the lobster industry has been faced with boycotts and has launched lawsuits to counter campaigns by conservation groups who are telling consumers that the lobster fishery is bad for endangered right whales. Geoff says the fishery needs to do a better job explaining to the public how it's trying to protect the whales, including shutting down fishing in areas whenever whales are spotted.

"We're doing our best," he argues. "These measures are world-class. Nobody else closes three thousand square kilometres of fishing over a single whale sighting. They have planes flying over throughout the season. When they're closed, everyone has to live with it. Nobody else does that. And it's worked, by and large. There's been a few entanglements. But here's the thing. I mean, we have to fish lobster. We have to give the economic value for these hundreds of communities. So it's a real balance."

We should expect that the centre point of the industry will continue to move northward. Grand Manan, once in the beating heart of Canada's lobster boom, is already on the wrong side of this geographical shift. This means seafood companies in Newfoundland and Labrador, Quebec, northern New Brunswick, and Cape Breton will become more dominant in the industry, while the importance of the fishery to New England will fade. As the catches shift, lobster exporters, shippers, and wholesale buyers will have to adjust where they source their shellfish from. Increasingly in the future, *American* lobster will be a *Canadian* product.

Farmed lobster may hold some promise in a world of prohibitively expensive wild-caught lobster. But although the science may have been proven, and a few companies are claiming early commercial-scale success, the long growth period of lobster and the cost of getting them to adulthood remains a challenge.

Others are trying to reduce the carbon footprint of the fishery. The Lobster Council of Canada and the Island Institute in Maine are among the organizations supporting the switch to electric marine motors, to wean the industry off its heavy reliance on diesel fuel. But there's a long way to go. Brad Small, the captain from Dipper Harbour who's been fishing since the 1970s, thinks battery technology has a limited future in an environment where corrosive seawater seems to get into everything.

But electric motors aren't what keeps Brad up at night. Nearly every day during lobster season, the boats go out from the harbour and return with holds that seem a little bit less full than they were a year before. He knows he's in a race against time. As he guides the *Small Fortune's* out for another day on the water, a big wave comes crashing over the bow, rocking the vessel violently. Brad braces himself against the wall, then begins to relax his body as the wave passes. He loves this life. But he wonders if his children will ever know what it was like, back when the traps always seemed to be full, and the sea seemed like it held an endless bounty.

"I just hope there's something worth leaving behind," he says.

Acknowledgements

This book started as an idea. Without the help and support of many people, it would have remained just that.

I owe so much to my editors at McClelland & Stewart and Penguin Random House Canada, especially Jenny Bradshaw, who patiently guided me through my first book project, and publisher Stephanie Sinclair, whose support for Canadian non-fiction is inspiring. I'm indebted to Shona Cook, Tonia Addison, and copy editor Shaun Oakey, whose sharp eye caught so many gaffes, typos, and repetition in the first draft.

I'd also like to thank my editors at the *Globe and Mail*, including but not limited to Greg McArthur, Renata D'Aliesio, Mark Iype, James Keller, Christine Brousseau, Sinclair Stewart, and David Walmsley, who encourage their journalists to think big and dig deeper on the stories we tell.

I'm grateful to Brad Small and the crew of the *Small Fortune's*, who took me to sea and allowed me to glimpse the life of a lobsterman. To Levi Paul Sr. and the Indigenous fishermen who helped me better understand the Mi'kmaq perspective on the lobster harvest. And to the many people who allowed me to sit down with them on the decks of their boats, in their pickup trucks, at kitchen tables, in restaurants, on processing floors, in airport hangars, and on countless wharfs to

ACKNOWLEDGEMENTS

show me how this business works. I can't do this kind of journalism without you.

I'm also indebted to the many archivists, librarians, and historians in towns large and small all around the East Coast who provided clippings, journals, photos, and other documents that helped bring this story to life. I want to acknowledge the important work of Joseph Gough and Jim Acheson, authorities on the history and politics of lobstering in both Canada and the U.S., and authors such as Mark Kurlansky, Elizabeth Townsend, Christopher White, Trevor Corson, Paul Greenberg, and Silver Donald Cameron, whose writings on lobster fishing culture and the seafood industry at large helped inform this book. I want to thank biologists Rick Wahle, Bryce Stewart, Rémy Rochette, Brian Beal, Riley Secor, Heather Koopman, and Andrew Westgate, who helped me better understand these incredible creatures.

I often relied on the work of many journalists in Canada, the U.S., and around the world, whose reporting on the lobster industry was an invaluable resource. That includes Tony Saxon, the talented editor and photographer who was the first to suggest this book should have used scratch-and-sniff technology.

I'd also like to thank my parents, who filled my childhood with books, newspapers, and magazines, and gave me curiosity about the world around me. My daughters, Norah and Suzy, for giving me a reason to keep going on this project, and their mother, Kate, for her support. And finally, my darling Hannah, for her endless encouragement, love, and positivity. I'm grateful to you all.

Notes

1: THE GREAT LOBSTER MACHINE

12   In 2011, Gateway shipped about 4,800 tonnes: Author interview with Karl Riches, Gateway Facilities, April 2023.

13   One of the first records: Henry Otis Thayer, *The Sagadahoc Colony: Comprising the Relation of a Voyage into New England* (Gorges Society, 1892), 43–45.

14   But a discovery made en route: Journal of Robert Davies, republished by Maine's First Ship, 2024.

14   "The boat," Captain Davies wrote: Journal of Robert Davies.

15   "The first person to eat a lobster . . .": Joseph Gough, *Managing Canada's Fisheries: From Early Days to the Year 2000* (McGill-Queen's University Press, 2006), 62.

15   Marine biologists who study these creatures: Author interview with Rémy Rochette, University of New Brunswick at Saint John, 2022.

16   There are other troubling signs: Mary Whitfill Roeloffs, "Harvest Down 5% amid Warming Ocean, Right Whale Regulations," *Forbes*, March 1, 2024.

16   Overall, catches have been trending downward: J.B. MacKinnon, "An Unnatural History of the McLobster," *The New Yorker*, September 12, 2015.

16   There are warnings in Canada, too: Nick Travis, "Season Starts as 'One We'd Like to Forget,' Says MFU," *Atlantic Fisherman*, September 10, 2024.

17   On the Magdalen Islands: Isabelle Larose, "'No smoke in the smokehouse' on the Magdalen Islands after ban on herring fisheries," *CBC News*, April 17, 2022.

17   "There are a lot of signs . . .": Author interview with Bryce Stewart, Department of Environment and Geography, University of York, October 2022.

18   In Canada, the total catch is twice: Fisheries and Oceans Canada's annual landings data.

20   "There is a portion of lobster from the commercial fishery . . ." author interview with Nat Richard, Lobster Processors of Canada, March 7, 2025.

2: THE LAST OF THE HUNTER-GATHERERS

22   On a good day, early in the season: Author interview with Brad Small, November 2023.

25   Down the coast, fishermen are talking: Patrick Whittle, "Maine Lobster Catch Dips to Lowest Levels Since 2009," Associated Press, March 1, 2024.

26   Most of the seafood consumed around the world: Paul Greenberg, *Four Fish: The Future of the Last Wild Food* (Penguin Books, 2010), 13–14.

27   People have fished out of Dipper Harbour: Ethel Anne Thompson, *The Tides of Discipline* (Print 'N Press Ltd., 1978), 43–45.

27   In the eighteenth and nineteenth centuries: Joseph Gough, *Managing Canada's Fisheries* (McGill-Queen's University Press, 2006), 37–84.

29   The problem is that North America's lobster industry: Livia Albeck-Ripka, "Climate Change Brought a Lobster Boom. Now It Could Cause a Bust," *New York Times*, June 21, 2018.

30   Lobster catches in the Bay of Fundy: Peter Lawton, *DFO Science Stock Status Report C3-61 (2001)*, Fisheries and Oceans Canada, October 2001, 1–3.

30   By January 2024, concern over the health of lobster stocks: Paul Withers, "Rogue Wave Hits Canadian Lobster Industry as U.S. Moves to Increase Minimum Legal Size," *CBC News*, January 19, 2024.

31   Almost as soon as Lobsterboys arrived: Chris Chase, "Bankrupt Lobster Boys Selling Canadian Plant, Employed Accountant Convicted of Misappropriating over USD 77 Million," *SeafoodSource*, May 7, 2024.

32   Down the road in Bay Shore: Connell Smith, "Chinese Demand Driving Investment in New Brunswick's Lobster Industry," *CBC News*, January 22, 2019.

3: A WHALE FOR THE KILLING

33   By 1775, half of the whaling vessels in Massachusetts: Nathaniel Philbrick, "How Nantucket Came to Be the Whaling Capital of the World," *Smithsonian Magazine*, December 2015.

34   By the early 1890s: Kristen Rogers, "A Whale Belonging to One of the Rarest Species Is 'Likely to Die,' After Entanglement, NOAA Says," *CNN*, January 18, 2023.

35   This long, migratory "superhighway": "Ships Speeding Through US 'Go Slow' Zones Meant to Protect Endangered Whales," Reuters, October 19, 2023.

35   The U.S. and Canadian governments have tried: *Approved Weak Inserts for the Atlantic Large Whale Take Reduction Plan*, NOAA Fisheries, October 28, 2024.

35   But whales are still getting entangled: "North Atlantic Right Whale Updates," NOAA Fisheries.

36   Others have gone into aquaculture: "The Nexus of Restoration and Economy for Local Aquaculture in Southeast New England," United States Environmental Protection Agency, June 21, 2024.

37   Although some fishermen: Rhythm Rathi, "N.B. Fishermen Test New Gear in Bid to Stay on the Water When Right Whales Spotted," *CBC News*, August 15, 2024.

38   The environmental group has been: Nicholas Lemann, "No People Allowed," *The New Yorker*, November 14, 1999.

39   Frustrated at being being kept off the water: Chris Chase, "Massachusetts Lobstermen Sue NOAA over Restricted Fishing Area," *SeafoodSource*, February 15, 2024.

41   Fishermen protest that this technology: Jordan Andrews, "Maine Lobstermen Fear Disaster as New Gear Regulations Take Effect," *Portland Press Herald*, May 1, 2022.

41   Nevertheless, both the American and Canadian governments: "Whalesafe Fishing Gear," Fisheries and Oceans Canada; "Developing Viable On-Demand Gear Systems," NOAA Fisheries.

41   Right whale regulations are coming: "Offshore Wind Research and Development," U.S. Department of Energy.

42   Three days before Christmas in 2022: Chris Chase, "US Lobster Fishery Could Get Longer Reprieve from Right Whale Rules," *SeafoodSource*, December 20, 2022.

4: THE BLEAKNESS

45   But there were plenty of warning signs: Nancy Balcom and Penelope Howell, *Responding to a Resource Disaster: American Lobsters in Long Island Sound, 1999–2004* (University of Connecticut, 2006), 5–10.

45   By August of 1999, fishermen began reporting: *1999 Report on Lobster Mortality in Long Island Sound* (Connecticut Department of Environmental Protection, 2000), 7.

45   By the end of the year: Richard A. French, *Assessment of Lobster Health in Long Island Sound 2000–2001* (New York Sea Grant Communications, 2001), 27.

45   This in the state that brought the world the lobster roll: Autumn Swiers, "Perry's Restaurant in Connecticut Is Where the Lobster Roll Was Invented," *Tasting Table*, March 17, 2024.

48   Long Island Sound is the southernmost: Nancy Balcom and Jack Pearce, "The 1999 Long Island Sound Lobster Mortality Event," *Journal of Shellfish Research* 24, no. 3 (2005): 691–97.

48   For years, warming waters seemed to be fuelling a boom in lobster populations: J.B. MacKinnon, "An Unnatural History of the McLobster," *The New Yorker*, September 12, 2015.

48   "The warmer water fuelled algal blooms . . .": Ibid.

49   So many lobsters were being unloaded: Ryan Breton, "Connecticut Lobster Industry Collapsing, Warming Water to Blame," *FOX61*, May 18, 2022.

49   Noank fishermen used to haul in half: Ben Rathbun, *The Lobster Hatchery* (Noank Historical Society, 2002), 1.

49   In 1898, a sudden die-off: Timothy C. Visel, "The Southern New England Lobster Fisheries Collapse of 1898–1905," The Sound School Regional Vocational Aquaculture Center, July 2015.

5: GRANDFATHER LOBSTERS

51   Thanks to an enzyme called telomerase: Maddy Massy-Westropp, "What Lobsters Can Teach Us About Immortality," University of New South Wales, September 19, 2022.

51   Accounts from fishermen: Joseph Gough, *Managing Canada's Fisheries* (McGill-Queen's University Press, 2006), 6.

52   For every fifty thousand eggs: "Fun Facts About Luscious Lobsters," NOAA Fisheries, July 2024.

52   When the eggs finally hatch: Ibid.

53   Lobster used to have a lowly place: Bayard Webster, "Primitive Lobster Revealed as One of Nature's Marvels," *New York Times*, October 12, 1982.

54   One lobster tagged in a study: "Fun Facts," NOAA Fisheries.

54   The lobster's strong sense of smell: Webster, "Primitive Lobster."

58   There are other warning signs: Jason S. Goldstein et al., "Recent Declines in American Lobster Fecundity in Southern New England: Drivers and Implications," *ICES Journal of Marine Science* 79, no. 5 (July 2022): 1662–74.

58   Not only are larger, super-productive females: Author interview with Rémy Rochette, University of New Brunswick at Saint John, December 7, 2022.

6: HEAT WAVES

61   The ten-day heat wave: Ed Kohn, "The Heat Wave of 1896 and the Rise of Roosevelt," *NPR*, August 10, 2011.

61   New York City in the late nineteenth century: Lewis Erenberg, *Steppin' Out: New York Nightlife and the Transformation of American Culture, 1890–1930* (Greenwood Press, 1981), 40.

62   Fisheries officials in Connecticut: Timothy C. Visel, "The Southern New England Lobster Fisheries Collapse of 1898–1905," The Sound School Regional Vocational Aquaculture Center, July 2015.

63   Researchers with the: Avery Siciliano et al., "Warming Seas, Falling Fortunes," Center for American Progress, September 10, 2018.

63   For fishing communities: Ibid.

63   As their traditional habitat: Author interview with Rick Wahle, School of Marine Sciences, University of Maine, August 2023.

65   In other studies, Heather attaches thermal loggers: Shane Fowler, "Tagging Along on the Secret Life of the Lobster," *CBC News*, December 13, 2022.

65   As seawater temperatures rise: Wahle, interview.

66   Especially troubling for biologists: Ibid.

7: SHELL DISEASE

69   First observed in lobster pounds: Stanley Cobb and Kathleen M. Castro, *Shell Disease in Lobsters: A Synthesis* (Rhode Island Sea Grant, May 2006).

70   It's a big reason: Ibid.

70   What scares scientists and fishermen: Christine Woodside, "Long Island Sound to Lobsters: Is This Farewell?," *Wrack Lines: A Connecticut Sea Grant Publication*, Fall/Winter 2018–19, 14–16.

71   Rhode Island's waters are warming: Barbara Moran, "Fishermen and Scientists Join Forces to Track Effects of Climate Change," *WBUR*, August 14, 2020.

71   It's common for water temperatures: Author interview with Mark Sweitzer, May 2023.

72   The federal government sponsored: Rhode Island Department of Environmental Management, *Draft Restoration Plan and Environmental Assessment for the January 19, 1996 North Cape Oil Spill*, September 14, 1998.

73   Over the past thirty years: Michael Carlowicz, "Watery Heatwave Cooks the Gulf of Maine," National Aeronautics and Space Administration, September 12, 2018.

75   Officials in Sweden used it: Patrick Whittle, "Shellfish Behavior: US Lobster Industry at Odds with Sweden," *Portland Press Herald*, April 12, 2016.

75   It's already showing up: Author interview with Rick Wahle, School of Marine Sciences, University of Maine, August 2023.

76   Research suggests that lobsters: Author interview with Riley Secor, University of Rhode Island, May 2023.

8: GOD'S WILL

78   There was still a gentle roll to the sea: Shannon Pittman, *Marine Transportation Safety Investigation M18A0078: Capsizing of the Fishing Vessel* Ocean Star II, Transportation Safety Board of Canada, May 12, 2018.

79   Glen helped Elijah up onto the boat: Author interview with Elijah Watts, July 2023.

79   Within minutes another fishing boat: Tom Ayers, "'He Just Kept the Place Alive': Port Hood Fire Chief Mourns Colleague's Death," *CBC News*, June 30, 2020.

80   Commercial fishing is one of: "Watchlist 2022: Commercial Fishing Safety," Transportation Safety Board of Canada.

80   Globally, there are more: FISH Safety Foundation, "More Than 100,000 Fishing-Related Deaths Occur Each Year, Study Finds," The Pew Charitable Trusts, November 3, 2022.

80   A deckhand in Canada is fourteen times more likely to die: Tavia Grant, "Sea Change," *Globe and Mail*, October 27, 2017.

80   In 2019, American fishermen: Commercial Fishing Incident Database (CFID), National Institute for Occupational Safety and Health (NIOSH), December 8, 2023.

81   "I feel like we're taking bigger and bigger risks . . .": Author interview with Gail Atkinson for the *Globe and Mail*, March 15, 2022.

82   Two days before she spoke: Michael Tutton and Keith Doucette, "Sister Confirms Death of Man Who Spent Five Hours in Waters off Eastern Nova Scotia," *Global News*, March 14, 2022.

82   When the *Ocean Star II* capsized: Pittman, *Marine Transportation Safety Investigation*.

9: DYING TO FISH

87   But no matter how hard he works: Author interview with Brad Small, Maces Bay, NB, November 2023.

89   Simmering for decades: Moira Donovan, "The Legal Fishery Sparking Arrests and Violence," *Hakai Magazine*, October 17, 2023.

89   Fishermen in this area are known: Nik DeCosta-Klipa, "This Strange, Lobster-Fueled Border Dispute off Maine Has Been Simmering Long Before Trump," Boston.com, July 22, 2018.

90   "Canadians are like Vikings . . .": Zane Schwartz, "The Tiny Islands Where Canada and America Are at War," *Maclean's*, July 22, 2015.

90   That was the Desautel decision: R v. Desautel, 2021 SCC 17, [2021] 1 S.C.R. 533.

90   The decision encouraged: Paige Williams, "Inside the Slimy, Smelly, Secretive World of Glass-Eel Fishing," *The New Yorker*, June 17, 2024.

91   On Brad's side of the bay: Andrew Bates, "Sitansisk First Nation and Chambook Residents in N.B. Concerned Lobster Facility Threatens 'Trees, Ocean and Quiet,'" *Telegraph-Journal*, February 2, 2024.

92   The scale of Canada's Indigenous lobster fleet: Paul Withers, "N.S. First Nations to Exercise Right to Moderate Livelihood During Upcoming Lobster Season," *CBC News*, November 20, 2023.

10: THERE WILL BE BLOOD

98   "You won't cut any more of our traps": Canadian Press, "Phillip Boudreau 'Murder for Lobster' Trial Hears from Accused," *CBC News*, November 25, 2014.

100   The morning he was killed: Silver Donald Cameron, *Blood in the Water: A True Story of Revenge in the Maritimes* (Viking Canada, 2020), 8–11.

101   In one particularly troubling case: Ian Urbina, "A Slaughter at Sea, a Grainy Video and Justice Delayed," *The Washington Post*, October 13, 2020.

101   China has the world's largest long-distance fishing fleet: *Blood and Water: Human Rights Abuse in the Global Seafood Industry*, Environmental Justice Foundation, June 2019.

101   In Grand Manan, New Brunswick: Calvin Trillin, "The House Across the Way," *The New Yorker*, June 18, 2007.

102   In the rich lobster grounds around Matinicus: Jim Flannery, "Lobster Wars: 'Good People' Doing Bad Things," *Soundings*, September 22, 2009.

103   Jim's work has often focused on the phenomenon: James M. Acheson, *The Lobster Gangs of Maine* (University Press of New England, 1988), 1–7.

103   In 2009, a festering dispute: Clarke Canfield, "Matinicus Shooting a Sore Subject," *Portland Press Herald*, July 21, 2010.

104   "They become the judge and jury . . .": Flannery, "Lobster Wars."

104   In 2016, another trap war erupted: Meredith Goad, "Lobstering Turf War off Maine Coast Brings Offer of \$15,000 Reward," *Portland Press Herald*, November 16, 2016.

104   "The idea is to make sure that people . . .": Abby Goodnough, "In Maine, Tensions over Ailing Lobster Industry," *New York Times*, August 22, 2009.

105   "If all fishermen everywhere . . .": Trevor Corson, "Learning from Maine's Lobster Wars," *The Atlantic*, September 16, 2009.

105   Maine approved a two-mile: Goodnough, "Tensions over Lobster Industry."

11: THE RIGHT TO FISH

107   The celebrations soon gave way: Keith Doucette, "Non-Natives Removing Lobster Traps Set by Indigenous Fishermen in Western N.S.," Canadian Press, September 20, 2020.

108   "Being out here, we feel at home . . .": Author interview with Levi Paul Sr., Saulnierville, N.S., September 2020.

109   Catching lobster under treaty rights: Nic Meloney, "Sipekne'katik First Nation Sues Nova Scotia over Restrictions on Buying Mi'kmaw Lobster," *CBC News*, February 3, 2021.

110   Donald Marshall, a member of Cape Breton's: *The Canadian Encyclopedia*, "Donald Marshall Jr.," by Edward Butts, August 6, 2009.

111   "The whole thing was orchestrated . . .": Author interview with Daniel Paul, Mi'kmaq historian, September 2020.

112   "He used to have night terrors . . .": Author interview with Colleen D'Orsay, September 2020.

113   Jason Marr was nearly done unloading: Author interview with Jason Marr, September 2020.

114   Nearly 68 per cent of Sipekne'katik families: Greg Mercer, "Two Decades After Landmark Ruling, Mi'kmaq Still Battling Prejudice over Fishing Rights," *Globe and Mail*, September 26, 2020.

12: LEGAL FIGHTS

116   The Oka Crisis: *The Canadian Encyclopedia*, "Kanesatake Resistance (Oka Crisis)," by Tabitha de Bruin, July 11, 2013.

118   First called Indian Brook: "History," Sipekne'katik First Nation, July 2023.

118   The residents of Sipekne'katik: "Shubenacadie (St. Anne's Convent)," National Centre for Truth and Reconciliation Archives.

118   "A lot of these people . . .": Author interview with Michael McDonald, January 2023.

119   "It is agreed that the said Tribe of Indians . . .": 1752 Peace and Friendship Treaty Between His Majesty the King and the Jean Baptiste Cope, transcribed from *R. v. Simon*, Supreme Court of Canada, 1985, Government of Canada.

119   Indigenous scholars have long argued: D. Bruce Clarke et al., *The Mi'kmaq Treaty Handbook* (Native Communications Society of Nova Scotia, 1987).

120   It wasn't until the Supreme Court's: *The Canadian Encyclopedia*, "The Marshall Case," by Heather Conn, April 11, 2020.

122   "Anybody who knows anything in this industry . . ." author interview with Nat Richard, Lobster Processors of Canada, March 7, 2025.

122   "What we have are just a bunch of fisheries officers . . .": McDonald, interview.

123   In 1927, Gabriel was convicted of hunting muskrats: "King vs. Sylliboy, 1928," Nova Scotia Archives, January 1, 2024.

123   In 2017, ninety after years after his conviction: Joan Weeks, "9 Decades After Hunting Conviction, Mi'kmaq Leader Gets Posthumous Pardon," *CBC News*, February 16, 2017.

123   "It's even kind of hard to talk about . . .": Maureen Googoo, "Sipekne'katik Fisherman Says Delay in Fishery Launch the Smart Decision for Now," *Ku'ku'kwes News*, June 5, 2021.

13: HISTORY REPEATING

125   But shortly after the Marshall decision: "Shots Fired in Burnt Church Fishing Dispute," *CBC News*, September 16, 2021.

126   "Everyone was concerned . . .": Greg Mercer, "Two Decades After the Burnt Church Crisis, Disputes Flare Up over Indigenous Fishing Rights in Atlantic Canada," *Globe and Mail*, September 27, 2020.

127   "History will show this present injustice . . .": August Lloyd, "Statement from Hereditary Chief of the Mi'kmaq Grand Council," New Brunswick Environment Network, February 20, 2013.

127   "We don't need governance . . .": Mercer, "Burnt Church Crisis."

128   "I don't think we should be surprised . . .": Ibid.

130   Between 2000 and 2007, Ottawa spent $354 million: Senate Standing Committee on Fisheries and Oceans, "Examining the Implementation of Indigenous Commercial Fishing Rights," briefing for the Minister of Fisheries and Oceans for her June 15, 2021, appearance at DFO.

131   "Despite our rights being affirmed . . .": Mercer, "Burnt Church Crisis."

131   Like other people in his community: Greg Mercer, "Four Men Arrested After Shots Fired at Indigenous Fisherman, RCMP say," *Globe and Mail*, December 14, 2020.

133   Older Mi'kmaq say the reaction: Greg Mercer, "Lobster Dispute, Frustration with Ottawa Could Turn the Tide on the Liberals in Nova Scotia," *Globe and Mail*, September 11, 2021.

14: A VOW TO FIGHT

137   "I was dragged into it kicking and screaming . . .": Author interview with Colin Sproul, February 2024.

138   One of the most famous: Geoffrey Wolff, *The Hard Way Around: The Passages of Joshua Slocum* (Knopf, 2010), 70–111.

139   His worry that rights-based access: Sarah Shamim, "Why are New Zealand's Maori Protesting over Colonial-Era Treaty Bill?," *Al Jazeera*, November 19, 2024.

139   Since then, Moana New Zealand: Will Trafford, "Sealord to Buy Independent Fisheries, Become New Zealand's Biggest Seafood Business," *Te Ao Māori News*, September 18, 2023.

139   He points to the Clearwater deal: Paul Withers, "First Nations Partner

with B.C. Company in $1B Purchase of Clearwater Seafoods," *CBC News*, November 9, 2020.

140 "I believe in the spirit of reconciliation . . .": Sproul, interview.

15: THE LOBSTER CARTEL

141 When the Riverside Lobster plant closed: Cliff White, "Champlain Seafood Closes Riverside Lobster in Nova Scotia," *SeafoodSource*, February 15, 2024.

141 "The lobster processing industry in Atlantic Canada . . .": Rachelle Gagnon, "Champlain Seafood Blames Lack of Lobsters for Permanent Closure of Meteghan, N.S., Processing Plant," *CBC News*, February 14, 2024.

141 Indeed, anecdotal reports: Sean Mott, "Some N.S. Lobster Fishers Report Serious Decline in Catches," *CTV News*, September 10, 2024.

142 Between 2017 and 2021: Claire Canet to House of Commons Standing Committee on Fisheries and Oceans, 43rd Parliament, 2nd Session, No. 37, June 16, 2021, Open Parliament.

143 In Newfoundland, a Dutch Crown corporation: Keith Sullivan, "Profit over People: Royal Greenland Isn't Here to Help Newfoundlanders," FFAW-Unifor, August 16, 2022.

143 "This allows the processors . . .": Author interview with Colin Sproul, February 2024.

144 Most famously, after four thousand fishermen: Molly Benjamin, "Remembering the Day Maine Fishermen Put Lobsters in a 'Bank,'" *Cape Cod Times*, March 26, 1998.

144 "What 'evil' had the fishermen done . . .": Ibid.

144 "In recent years, mom-and-pop . . .": Martin Mallet to House of Commons Standing Committee on Fisheries and Oceans, 43rd Parliament, 2nd Session, No. 37, June 16, 2021, Open Parliament.

145 "Times are changing . . .": Author interview with Martin Mallet, October 10, 2023.

146 Colin calls the corporate concentration: Sproul, interview.

147 "If we push the prices up too high . . .": Author interview with Geoff Irvine, Lobster Council of Canada, August 14, 2023.

149 He points out that Clearwater: Tom Ayers, "Nova Scotia lobster industry facing headwinds after Clearwater exit from live shipping," *CBC News*, February 28, 2025.

16: CHINESE LOBSTER

154 "It's crazy . . .": Author interview with Alex Puig Coll, October 26, 2023.

155 China began to loosen the rules: Yeling Tan, "How the WTO Changed China: The Mixed Legacy of Economic Engagement," *Foreign Affairs*, March/April 2021.

155 It wasn't until American exporters: Author interview with Peter Redmayne, October 27, 2023.

156 As China opened up: Paul Withers, "China Is Close to Becoming Canada's Largest Export Market for Live Lobster," *CBC News*, January 7, 2020.

156 "They started serving it in high-end . . .": Redmayne, interview.

157 Global exports of lobster increased: Food and Agriculture Organization of the United Nations, "Reduced US Landings Will Open New Opportunities for Canadian Suppliers," *FAO Globefish*, October 25, 2023.

157 When U.S. President Donald Trump: Michael Gorman, "Canadian lobster industry officials say U.S. tariff threat cause for even more diversified markets," CBC News, January 21, 2025.

158 "It has never been more important . . ." TJ Colello, "Cape Breton seafood exporters wary of potential U.S. tariffs," *Cape Breton Post*, February 5, 2025.

159 In 1998, lobstermen in Maine were getting $2.92 a pound: "Historical Maine Lobster Landings," Maine Department of Marine Resources, https://www.maine.gov/dmr/sites/maine.gov.dmr/files/docs/lobster.table.pdf.

160 "China is still a frustrating market . . .": Redmayne, interview.

160 China's appetite for the world's seafood: Cliff White, "US, Canadian Lobster Exports to China Set to Take Off in 2023," *SeafoodSource*, January 19, 2023.

161 But politics sometimes gets in the way: "US Lobster Exports to China Fall Dramatically amid Tariffs," Associated Press, August 26, 2019.

162 By the end of 2024, China was importing: Raechel Huizinga, "China's looming seafood tariffs just add to 'craziness,' says lobster organization," CBC News, March 10, 2025.

162 "Increasingly, our focus has been China . . .": Greg Mercer, "The U.S.–China Trade War Is a Boon for Atlantic Canada's Lobster Harvesters. But What's the Catch?," *Globe and Mail*, November 28, 2019.

162 Australia was caught in a prolonged trade battle: Paul Karp and Helen Davidson, "China Bristles at Australia's Call for Investigation into Coronavirus Origin," *The Guardian*, April 29, 2020.

17: THE YELLOW SEA

165 It's estimated that about 40 per cent: "Losing Tidal Flats Around the Yellow Sea," NASA Landsat Science, May 28, 2014.

165 "Before, it was a rich man's food . . .": Author interview with Ziyi Ye, October 27, 2023.

165 In 2022, China imported $5.65 billion: Louis Harkell, "Trade Insights: China's Frozen Shrimp Imports Pass $5bn in 2022," *Undercurrent News*, December 23, 2022.

166 "Lobster is especially popular for weddings . . .": Author interview with Ji Peng, October 26, 2023.

167 Until 2017, Bayshore Lobster and Seafood: Connell Smith, "Chinese Demand Driving Investment in New Brunswick's Lobster Industry," *CBC News*, January 22, 2019.

167 "Most of our growth has depended . . .": Author interview with Elley Chen, October 28, 2023.

169 But companies with Chinese origins: Paul Withers, "Nova Scotia MP Questions Chinese 'Control' over Lobster Industry," *CBC News*, May 25, 2023.

170 "Young, wealthy Chinese people . . .": Author interview with Geoff Irvine, Lobster Council of Canada, August 14, 2023.

18: THE MUSIC OF THE SEA

173 In the 1870s, two decades: Ronan Browne, "History of the Irish Lobster," Trinity College Dublin, December 2001.

174 The island's lobster harvest peaked: "Developing the Irish Seafood Industry," Bord Iascaigh Mhara annual report, 2021.

174 "I don't know if there will be a fishery left . . .": Author interview with Gerry Sweeney, October 23, 2022.

175 Though the rules around Fish Fridays: Maria Godoy, "Lust, Lies and Empire: The Fishy Tale Behind Eating Fish on Friday," *NPR*, April 6, 2012.

175 And the tradition spawned innovations: Paul Clark, "Filet-O-Fish Inventor Brought Patrons Back to McDonald's," *Cincinnati Enquirer*, February 26, 2016.

175 "That's the kernel of the problem . . .": Author interview with Seamus Breathnach, October 20, 2020.

176 His grandfather was a lobster fisherman: Noel Campbell, "Lobster Fishing Season Is No Pot Luck," National Museum of Ireland, August 6, 2022.

176 Between 1845 and 1855: "Irish-Catholic Immigration to America," Library of Congress.

177 Carna's population before the Great Famine: Gerard Moran, "Uncovering 'the Forgotten Famine' of 1879–81 in the West of Ireland," *Journal of the Galway Archaeological and Historical Society* 72 (2020): 83–94.

177 The Irish may have adopted modern: Clodagh Kilcoyne, "Seafarers' Pilgrimage to MacDara's Island," Reuters, July 21, 2016.

178 It's believed the saint built: Dan MacCarthy, "Islands of Ireland: If You Ever Posted a Letter in the 1980s You'll Be Familiar with Macdara's Island," *Irish Examiner*, July 8, 2022.

179 Patsy Mullins, who has been: Lorna Siggins, "Alarm in Irish Lobster Sector over Hogan's New US Trade Deal," *Irish Independent*, August 29, 2020.

19: THE LOBSTER CAPITAL OF EUROPE

181 The lobster fishery on England's scenic east coast: Rajeev Syal, "Yorkshire Lobster Exporter Says Brexit Costs Have Forced It to Close," *The Guardian*, February 28, 2021.

182 "We don't do surf-and-turf here . . .": Author interview with Julie Hill, October 28, 2023.

182 Although archeological digs have found: Elisabeth Townsend, *Lobster: A Global History* (Reaktion Books, 2011), 24–25.

183 A 2019 Norwegian study suggests: Alf Ring Kleiven et al., "Technological Creep Masks Continued Decline in a Lobster (*Homarus gammarus*) Fishery over a Century," *Scientific Reports* 12, no. 1 (February 28, 2022): 3318.

183 The Norwegians began exporting live lobster: Townsend, *Lobster*, 43.

184 "There's lots going on, and we don't . . .": Author interview with Jamie Robertson, October 25, 2022.

184 For decades, British fishermen: Ibid.

184 The traditional Yorkshire coble: "Yorkshire's Last Traditional Sailing Fishing Coble," Scarborough Maritime Heritage Centre.

185 The French have long loved their "Bretagne" lobster: Townsend, *Lobster*, 38–39.

185 And yet limits on its harvest are almost nonexistent: Author interview with Bryce Stewart, October 18, 2022.

187 Some conservation measures are slowly being introduced: S. Coulthard and L. Barnes, "Fisher-Led Perspectives on Crab and Lobster Management in

Northumberland," Newcastle University report for the Northumberland Inshore Fisheries and Conservation Authority (NIFCA) and Blue Marine Foundation, July 2023.

20: THE PRINCE OF BRITTANY

189 "And then there are the prices . . .": Nicholas Lander, "Guy Savoy, One of Paris's Most Luxurious Restaurants," *Financial Times*, August 23, 2018.

189 Since the Middle Ages: Elisabeth Townsend, *Lobster: A Global History* (Reaktion Books, 2011), 27–28.

189 And of course it was a French head servant: Daniel M. Lavery, "Good Servants and Bad Masters," *Lapham's Quarterly*, August 29, 2022.

189 Across the Atlantic, however, French chefs: Stephane Bern, "Le homard à l'Américaine," *Europe1*, November 25, 2020.

190 Two centuries earlier, François Pierre de la Varenne: Townsend, *Lobster*, 38–39.

190 Upper-class French began eating lobster: Ibid., 26.

190 The most well-known French contribution: Erica Martinez, "The Unexpected Origin Story of Lobster Thermidor," *Food Republic*, February 15, 2024.

191 It was an eminent French zoologist: "Henri Milne Edwards," *Proceedings of the American Academy of Arts and Sciences* 21 (May 1885–May 1886): 547.

191 When they began fishing for spiny lobsters: Anne-Sophie Grollemund, "The Lobster War," U.K. National Archives, February 21, 2019.

191 "The attitude of France is inadmissible . . .": Ibid.

192 Cédric Delacour, a bearded fisherman: Emily Monaco, "France's Illustrious Blue 'Breton Lobster,'" *BBC*, July 14, 2023.

192 "It's a sign that the species feels endangered . . .": Ibid.

192 Warming waters aren't the only sign: "Foreign Trade in Fishery and Aquaculture Products," FranceAgriMer annual report, 2010.

193 Meanwhile, warming water off the coast: Hannah Thompson, "Octopus Invasion Threatens Lobster and Livelihoods in Brittany," *The Connexion*, August 29, 2021.

193 One seafood auction house in Finistère: Sylvain Moreau, "Mysterious Invasion of Octopuses in Brittany: Is It So Worrying?," *GEO*, April 19, 2023.

21: "I'M STILL HERE"

195   In the late nineteenth century, the Underwood fish packing plant: Penobscot Marine Museum, "Sardine Factory at Jonesport, ca. 1910," Maine Historical Society.

196   In that era, Jonesport was home to a major shipyard: "A Town for Whom the Sea Is Life," Town of Jonesport.

196   In 1960, the last sardine cannery here closed: Laurie Schreiber, "Herring, Once King on Coast, Now Worries Regulators," Island Institute, October 1, 2018.

196   "We'd get bumped every now and then . . .": Author interview with Bert Sidney, October 17, 2022.

197   Rising water temperatures in the Gulf of Maine: Penelope Overton, "A Fast-Warming Gulf of Maine Is Rising Faster than Ever," *Portland Press Herald*, June 7, 2024.

197   Sid wonders if the restocking programs: Frank Nicosia and Kari Lavalli, "Homarid Lobster Hatcheries: Their History and Role in Research, Management, and Aquaculture," NOAA Scientific Publications Office, *Marine Fisheries Review* 61, no. 2 (1999): 1–57, http://hdl.handle.net/1834/26407.

197   "I remember the first time I saw a shipment . . .": Sidney, interview.

197   The state's Department of Marine Resources: K.F. Drinkwater and B. Petrie, "A Note on the Long-Term Sea Surface Temperature Records at Boothbay Harbor, Maine," *Journal of Northwest Atlantic Fishery Science* 43 (2011): 93–101.

198   Biologists generally agree that once ocean temperatures rise: Erin Koenig, "American Lobsters Feeling the Heat in the Northwest Atlantic," *GeoSpace* (blog), January 22, 2018.

198   Biologists have been warning for a while: Arnault Le Bris, "New Study: Warming, Conservation, and Lobsters," Gulf of Maine Research Institute, December 31, 2018.

199   "The more I grow my operation . . .": Ben Speggen, "Tomorrow's Lobsterman, Today: A Look into the Future from Eastport," *Craftsmanship Quarterly*, January 7, 2023.

199   Climate change isn't just reducing catches: Nick Perry, "Record High Tide in Maine Washes Away 3 Historic Fishing Shacks," *The Boston Globe*, January 14, 2024.

199 When Hurricane Fiona tore through Atlantic Canada: "Hurricane Fiona Causes $660 Million in Insured Damage," Insurance Bureau of Canada, October 19, 2022.

200 But Maine has been through this before: Schreiber, "Herring, Once King."

22: LOBSTER ON A ROLL

202 "When we first started back in the late 80s . . .": Author interview with Marc Worrall, December 3, 2023.

203 "My real concern now is the resource of lobsters . . .": Ibid.

204 Ironically, it was a French restaurant chain: Dustin Wlodkowski, "A Lobster Roll from Paris Lands in Portland," *News Center Maine*, July 7, 2018.

204 He told the Paris newspaper *Le Figaro*: Alice Bosio, "The Best Lobster Roll in the World Can Be Tasted in Paris," *Le Figaro*, July 9, 2018.

204 "I think we have a lot more market . . .": Author interview with Luke Holden, December 18, 2023.

204 Luke's Lobster was born out of his frustration: Janet Morrissey, "A Restaurant's Sales Pitch: Know Your Lobster," *New York Times*, August 24, 2016.

205 "Our partner understood that folks in Tokyo . . .": Author interview with Ben Conniff, December 18, 2023.

206 The first year they opened, 2009, they were paying: Ibid.

23: THE END OF MCLOBSTER

208 The idea for a low-cost, mass-produced lobster roll: SaltWire Network, "Created in Amherst, McLobster Has Gone On to a Bigger Stage," *PNI Atlantic News*, May 17, 2016.

208 "People were talking about what we could offer . . .": Author interview with Danny Moore, January 20, 2023.

208 In Maine, in the early 1990s: "Historical Maine Lobster Landings," Maine Department of Marine Resources, https://www.maine.gov/dmr/sites/maine.gov.dmr/files/docs/lobster.table.pdf.

208 "The root cause of lobster's slow migration . . .": J.B. MacKinnon, "An Unnatural History of the McLobster," *The New Yorker*, September 12, 2015.

209 Back in 1992, the McLobster sandwich was his father's idea: Moore, interview.

209 He'd been one of a group of restaurant managers: "When a Canadian Brought McDonald's Fast Food to Moscow," *CBC Archives*, January 13, 2019.

210   The campaign to expand into Russia: Sam Roberts, "George Cohon, Who Brought Big Macs to Moscow, Dies at 86," *New York Times*, December 12, 2023.

211   "We'd see the buses pull in . . .": Moore, interview.

211   "It was an expensive inventory item . . .": Ibid.

24: BOOM TIMES

216   Fishermen in Newfoundland and Labrador: "Seafood Industry Year in Review (2022)," Government of Newfoundland and Labrador, 15, https://www.gov.nl.ca/ffa/files/Seafood-Industry-Year-in-Review-2022.pdf.

216   Lobster has quickly become one of the most important fisheries: Jane Adey, "Land & Sea Heads to the Lobster Grounds to Find Out Why the Fishery Is Booming," *CBC News*, November 23, 2024.

216   "Science was right, and the old fishermen . . .": Author interview with Rick Crane, July 14, 2023.

216   That was never the intention: Jenny Higgins, "Economic Impacts of the Cod Moratorium," Newfoundland and Labrador Heritage Website, 2008.

217   The commercial extinction of northern cod: Michela Rosano, "Cod Moratorium: How Newfoundland's Cod Industry Disappeared Overnight," *Canadian Geographic*, July 11, 2022.

217   "The only thing you could do to prevent starvation . . .": Crane, interview.

218   The fishery was essentially the Wild West: "American Lobster – Lobster Fishing Area 3-14C," Fisheries and Oceans Canada, September 9, 2021.

219   Newfoundland's new lobster boom: Ibid.

220   In the spring of 2023, Rick opened: Colleen Connors, "Fisherman Shells Out $500K for Lobster Holding Tank to Maintain Control over Selling Process," *CBC News*, June 7, 2023.

220   In 1999, when he first started fishing: Crane, interview.

221   In September 2022, Newfoundland's southwestern coast: Sarah Smellie, "A Year After Fiona, a Traumatized Newfoundland Town Backs Away from the Sea," Canadian Press, September 24, 2023.

25: THE TOWN THAT LOBSTER SAVED

222   Tignish, a sleepy, tree-lined community: "Population and Dwelling Counts: Canada, Provinces and Territories, and Census Subdivisions (Municipalities), Prince Edward Island," Statistics Canada, February 9, 2022.

222　The Filipinos are here because of Royal Star Foods: "A Dream of Canada Comes True with the Help of a P.E.I. Lobster Plant," *CBC News*, May 2, 2018.

223　"There's no houses with the lights turned off anymore . . .": Author interview with Francis Morrissey, March 20, 2023.

223　The story of lobster in fishing communities: *Report on* Homarus americanus *Landings Canada USA 2016–2021*, Lobster Council of Canada.

224　To the collapse of the groundfish sector: "Groundfish – Gulf of St. Lawrence (NAFO) Subdivisions 3Pn, 4Vn and Divisions 4RST," Fisheries and Oceans Canada, January 2017.

225　Part of Royal Star's success: Eric McCarthy, "Royal Star Foods Getting Help with Cold Storage Expansion," *The Guardian*, October 29, 2013.

225　In Qatar, the British-owned Burger & Lobster: Menu from Burger & Lobster, The Pearl Island, 6 La Croisette, Doha, Qatar.

226　The global demand for Atlantic lobster: Alexander Chafe, "Tignish Fisheries Cooperative Association, Supporting Fishers for 99 Years and Counting," *Atlantic Business*, October 18, 2024.

226　One of the Tignish co-op's founders: Sue Calhoun, *A Word to Say: The Story of the Maritime Fishermen's Union* (Nimbus Publishing, 1991), 274.

26: WELL-TRAVELLED LOBSTER

228　But William Lyman Underwood: Genevieve Wanucha, "Two Happy Clams," *MIT Technology Review*, February 24, 2009.

228　The explosion of the canning industry: Joseph Gough, *Managing Canada's Fisheries* (McGill-Queen's University Press, 2006), 48–63.

228　"Now the rapid development of lobster . . .": Ibid.

229　Underwood became a household name: Andrew Smith, *Eating History: Thirty Turning Points in the Making of American Cuisine* (Columbia University Press, 2009).

229　In 1810, Appert won a prize: Stephen Schaber, "Why Napoleon Offered a Prize for Inventing Canned Food," *Planet Money* (NPR blog), March 5, 2012.

230　"The contents of such cans were found . . .": Lyman Underwood, *The Technology Quarterly and Proceedings of the Society of Arts*, Massachusetts Institute of Technology, Volume X, 1896.

230　Helped by MIT food scientist: Anne Trafton, "Canned, Good," *MIT News*, January 12, 2011.

230 "It used to be there were canneries . . .": Author interview with Bernie MacDonald, January 16, 2023.

231 Canneries were eventually replaced: Gough, *Managing Canada's Fisheries*, 185–200.

231 "There's a lot of 'Greenpeace people' out there . . .": Author interview with Casey Benson, August 23, 2022.

232 "When I first started out, the furthest any of our lobster . . .": Author interview with Morton Benson, August 23, 2022.

233 Livingston Stone, a Harvard-educated federal fish commissioner: Lawrence P. Gooley, "Livingston Stone: Leading 19th Century Fisheries Expert," *Adirondack Almanack*, February 23, 2016.

233 One of the earliest attempts came in 1814: Spencer Baird, *United States Commission of Fish and Fisheries, Report of the Commissioner*, 1889.

233 In 1874, Livingston, then the deputy fish commissioner: M.L. Perrin, "Transportation of Lobsters to California – 1874," republished in *Fishermen's Voice* 16, no, 7, July 2011.

234 A small percentage of all wild lobster have gaffkemia: Susan Bower, "Synopsis of Infectious Diseases and Parasites of Commercially Exploited Shellfish: Gaffkemia of Lobsters," Fisheries and Oceans Canada, 2007.

234 "After the fifth day crowds of lobsters . . .": Perrin, "Transportation of Lobsters."

236 "In part these changes [to food production] were brought . . .": R.J. Ghelardi, "Oceanographic Phase of the Fatty Basin Study for a Lobster Transplant," Fisheries Research Board of Canada, 1967.

27: FISHERMEN FARMERS

238 "I was sixty-something years old . . .": Author interview with Bob Baines, August 6, 2023.

239 Bob uses his knowledge as an experienced fisherman: Gabriella Gershenson, "'I'm Not a Quitter': Lobstermen Turn to Kelp Farming in the Face of Climate Crisis," *The Guardian*, May 19, 2020.

239 Lobstermen in the state caught more: Bloomberg News, "Lobster Prices Jump, but Diners Pay Up," *National Fisherman*, December 21, 2016.

239 A typical kelp farm can bring in between: Gershenson, "'I'm Not a Quitter.'"

240 But it's not just anecdotal evidence: Jia-Rui Shi et al., "Sea Surface Temperature

Research Provides Clear Evidence of Human-Caused Climate Change," Woods Hole Oceanographic Institution, March 19, 2024.

240   "You can just see the increase in the water . . .": Baines, interview.

240   "We have so much fishing effort now . . .": Ibid.

241   As the catches grew along the New England coast: Kenneth Martin and Nathan Lipfert, *Lobstering and the Maine Coast* (Maine Maritime Museum, 1985).

241   For nearly a century, lobster catches: Author interview with Rick Wahle, August 10, 2023.

241   The boom had been decades in the making: Ibid.

241   This marked increase in effort makes little sense: Jim Acheson, "Confounding the Goals of Management: Response of the Maine Lobster Industry to a Trap Limit," *North American Journal of Fisheries Management* 21, no. 2 (May 2001): 404–16.

242   "This escalation makes no sense . . .": Jim Acheson, *The Lobster Gangs of Maine* (University Press of New England, 1988).

242   "My nephew went back and got his captain's license . . .": Baines, interview.

243   When *The New York Times*' Jess Bidgood: Jess Bidgood, "A Fisherman Tries Farming," *New York Times*, October 10, 2017.

243   "Climate change really helped us . . .": Livia Albeck-Ripka, "Climate Change Brought a Lobster Boom. Now It Could Cause a Bust," *New York Times*, June 21, 2018.

243   In 2012, Maine's fishermen realized: Bill Trotter, "2012 Maine Lobster Catch Increases by 18 Million Pounds, but Price Continues to Decline," *Bangor Daily News*, January 4, 2013.

243   "That year, we had an anomalously warm winter . . .": Author interview with Sam Belknap, August 18, 2023.

244   "If your business model is based on a million-dollar loan . . .": Ibid.

245   If that happens, more of Maine's coastline: Jacqueline Weaver, "Islands Were First Footholds for European Settlers," Island Institute, 2020.

28: LOBSTER IN THE LIBRARY

247   But he needed space to grow his lobster babies . . .": Author interview with Brian Beal, August 9, 2023.

248   "We're a fishing community, so people were all for it . . .": Author interview with Cynthia Rowden, August 21, 2023.

249 "I mean, when the female flips its tail . . .": Beal, interview.

250 Between 1986 and 1992, Brian and his research assistants: Brian Beal and Samuel Chapman, "Methods for Mass Rearing Stages I–IV Larvae of the American Lobster," *Journal of Shellfish Research* 20, no. 1 (2001): 337–46.

250 The first hatcheries began in New England and Atlantic Canada: Joseph Gough, *Managing Canada's Fisheries* (McGill-Queen's University Press, 2006), 197–225.

250 The St. Andrews Lobster Culture Laboratory: Gregor Reid, ed., *The History of Aquaculture Research and Training in St. Andrews, New Brunswick (Bulletin of the Aquaculture Association of Canada* 110, no. 1 [2012]).

251 "All of them had to do with the same thing . . .": Beal, interview.

251 Plenty of others have tried to meet the challenge: Gladwin Hill, "Lobster Farming Called Promising," *New York Times*, June 9, 1975.

252 Some continue to bet heavily on farmed lobster: Asbjørn Drengstig and Asbjørn Bergheim, "First Commercial Lobster Culture Module Established in Norway," Global Seafood Alliance, May 1, 2007.

252 Rick Stein, a British TV personality and celebrity chef: "Land Ahoy for 1m Pound Lobster Farming," *FarmingUK*, November 1, 2008.

252 The Norwegians, accustomed to commercial scale salmon farming: Dominic Welling, "Norwegian Lobster Farm 'Bouncing Back,'" *IntraFish*, September 25, 2014.

253 In 2012, Darden Restaurants: Dina Spector, "Red Lobster's Owner Is Taking a Big Step Toward Complete Global Lobster Domination," *Business Insider*, February 25, 2013.

254 "It's too difficult for our climate . . .": Author interview with Robin Zhou, October 28, 2023.

254 "I think it would just end up fairly low down . . .": Author interview with Gregor Reid, April 4, 2024.

255 The seafood business already understands: Alex Mayyasi, "The Invention of the Chilean Sea Bass," *Priceonomics*, April 28, 2014.

255 "But here's the deal . . .": Beal, interview.

29: LIFE AFTER THE BOOM

256 People have been collecting, drying and eating dulse: Doug Scott, "The Dulse Collectors," *Saltscapes Magazine*, December 2024.

256 Elton Greene, one of the most experienced dulsers: Greg Mercer, "On Grand Manan, Dulsers Endure, Despite Warming Ocean Waters," *Globe and Mail*, August 8, 2022.

257 "We didn't know where they were all coming from . . .": Author interview with Elton Greene, August 2, 2022.

257 "There was a joke on Grand Manan . . .": Author interview with Andrew Westgate, August 6, 2023.

259 "The numbers just weren't there . . .": Author interview with Brian Guptill, August 6, 2023.

259 In Maine, where eighteen thousand people depend on the lobster fishery: "The Future of Lobster," Island Institute, 2021.

260 "It needs to be shut down and turned into a conservation area . . .": Author interview with Howard Robbins, November 27, 2023.

261 U.S. regulators, concerned about warnings signs: Paul Withers, "Rogue Wave Hits Canadian Lobster Industry as U.S. Moves to Increase Minimum Legal Size," *CBC News*, January 19, 2024.

261 Dave Cousens, former long-time president: Laurie Schreiber, "David Cousens, MLA's Retiring Prez, Leaves a Great Legacy," *Fishermen's Voice*, 2018.

262 "We have to think about the world being a smaller place . . .": Author interview with Geoff Irvine, Lobster Council of Canada, August 14, 2023.

Index